Principles & Methods of Sampling in Animal Diseases Surveys

Principles & Methods of Sampling in Animal Diseases Surveys

Editor

Julio Mendes

scitus academics

Principles & Methods of Sampling in Animal Diseases Surveys

Edited by **Julio Mendes**

ISBN: 978-1-68117-665-9
Library of Congress Control Number: 2016947447

www.scitusacademics.com

Notice

Printed in United States of America on Acid Free Paper

Preface

Animal diseases can have a negative impact on animal welfare, public health and the economy. This particularly applies to infectious diseases like bird flu, Q fever and swine fever. Farmers, vets and the government work together to control these diseases. Animal diseases result in increased mortality and morbidity in livestock populations. Disease may affect performance through reduced fertility, delays in reaching maturity for reproduction or sale, decreased production of milk, eggs, or wool, decreased draught power, or decreased weight of fattening or cull animals. Animals raised by small-scale producers and backyard farmers in developing countries tend to be plagued with re-infection, and they typically lack access to diagnosis and control programs. Epidemiological studies usually involve sampling from livestock populations in some way in order to make inferences about a disease or diseases present in these populations.

Principles & Methods of Sampling in Animal Diseases Surveys presents methods and techniques for conducting an animal disease surveillance program, and developing an animal health monitoring system. It will be of valued in veterinary epidemiology and regulatory medicine, where there is need for a crisp assortment of material on animal disease monitoring, surveillance, and reporting tactics. This need arises from a new age of international trade regulations established on animal diseases, new demands for accountability in utilization of research funds, and calls for prioritizing and economically justifying animal health regulatory and diagnostic accomplishments. The book is intended to researchers & practitioners, and other animal health authorities who are working in world-wide based programs.

Table of Contents

CHAPTER 1

Evolution of equine infectious anaemia in naturally infected mules with different serological reactivity patterns prior and after immune suppression

Gian Luca Autorino, Claudia Eleni, Giuseppe Manna, Raffaele Frontoso, Roberto Nardini, Cristiano Cocumelli, Francesca Rosone, Andrea Caprioli, Lavinia Alfieri, Maria Teresa Scicluna*

Istituto Zooprofilattico Sperimentale Lazio e Toscana "M. Aleandri" Via Appia Nuova, 1411, 00178 Rome, Italy

ABSTRACT

Information on equine infectious anaemia (EIA) in mules, including those with an equivocal reaction in agar gel immunodiffusion test (AGIDT), is scarce. For this, a study was conducted to evaluate the clinical, viral loads and pathological findings of two groups of naturally infected asymptomatic mules, respectively with a negative/equivocal and positive AGIDT reactivity, which were subjected to pharmacological immune suppression (IS). A non-infected control was included in the study that remained negative during the observation period. Throughout the whole study, even repeated episodes of recrudescence of EIA were observed in 9 infected mules, independently from their AGIDT reactivity. These events were generally characterised by mild, transient alterations, typical of the EIA acute form represented by hyperthermia and thrombocytopenia, in concomitance with viral RNA (vRNA) peaks that were higher in the Post-IS period, reaching values similar to those of horses during the clinical acute phase of EIA. Total tissue viral nucleic acid loads were greatest in animals with the major vRNA activity and in particular in those with negative/equivocal AGIDT reactivity. vRNA replication levels were around 10–1000 times lower than those reported in horses, with the animals still presenting typical alterations of EIA reactivation. Macroscopic lesions were absent in all the infected animals while histological alterations were characterised by lymphomonocyte infiltrates and moderate hemosiderosis in the cytoplasm of macrophages. On the basis of the above results, even mules with an equivocal/negative AGIDT reaction may act as EIAV reservoirs. Moreover, such animals could escape detection due to the low AGIDT sensitivity and therefore contribute to the maintenance and spread of the infection.

KEYWORDS

Equine infectious anaemia; EIA; Mule; AGIDT; Immune suppression; Viral loads; Clinical signs; Pathological alterations

1. INTRODUCTION

Equine infectious anaemia (EIA) is a viral disease that affects all domestic members of the genus Equus spp. (horse, donkey and mule) and is characterised by intermittent fever, progressive anaemia, emaciation and death in severe cases. Clinical forms of EIA are acute, chronic and inapparent, with the latter being the most frequent. Animals in the chronic and inapparent phase may experience a recrudescence of the infection as a consequence of severe stress, hard work or the presence of other diseases (Quinlivan et al., 2007). EIA virus (EIAV), a Lentivirus of the Retroviridae family, is responsible for the infection that is characterised by a restricted tropism for the equine monocyte-macrophage lineage with productive replication occurring only in the differentiated tissue macrophages (Oaks et al., 1998).

During the surveillance programme conducted in Italy, between 2007 and 2012, that used the ELISA as a screening test and agar gel immunodiffusion test (AGIDT) as a confirmatory test, prevalence in mules was found to be significantly higher than in horses and donkeys

(Sala et al., 2012). In addition, equivocal AGIDT reactions were more frequent among this hybrid species (Scicluna et al., 2013).

Studies on the carrier status with negative and equivocal AGIDT reactions are respectively described for horses, by McConnell and Katada, (1981) and by Issel and Adams (1982), in which the authors succeeded in the transmission of EIAV from such animals. Similar investigations were also conducted to evaluate the correlation of the AGIDT serological reactivity pattern with plasma viral loads at different time intervals and in the tissues of experimentally infected horses (Harrold et al., 2000 and Hammond et al., 2000). In addition, in a study conducted by Cook et al. (2003), the authors report that the acute form of the disease in the horse is characterised by the typical EIA alterations of hyperthermia and thrombocytopenia which are observed when the plasma viral RNA (vRNA) copies exceed concentrations of $\log_{10}$ 7.7/ml, defined as the pathogenic threshold (Cook et al., 2001). Different was the evolution observed in donkeys as these animals maintained low viremic levels in the absence of clinical signs (Cook et al., 2001).

Clinical and virolgical studies on EIA in mules are limited to two publications, in which the authors, in one of them (Spyrou et al., 2003), describe in detail the evolution of the natural and experimental infection observed in two subjects for each condition, for a time period that was however insufficient to completely elucidate this. Also, the presence of EIAV in these animals was reported only qualitatively. Considering the relatively high EIA prevalence levels in the Italian mule population and their elevated density in restricted areas where they are used as working animals, the scientific literature on the importance of the role that they might play as EIAV reservoirs is to date scarce. In view of this, an observational study was conducted to assess the clinical, pathological and virological parameters of naturally EIA infected mules having equivocal/negative AGIDT reactions, according to the scoring system proposed by Issel et al. (1999). In addition, the results obtained from these animals were compared to those having a clearly positive AGIDT reaction. As all the study animals were presenting an inapparent form of the infection, they were subjected to a pharmacological immune suppression (IS) as described byCraigo et al. (2007) for the reactivation of the disease. Further to the serological findings, already described by Scicluna et al. (2013), this paper reports the clinical evolution of EIA, in relation to the plasma vRNA levels, with the aim to investigate the correlations among these characteristics and evaluate the results in relation to the risk such animals may represent in the epidemiology of EIA. Viral replication was assessed in terms of vRNA copies in plasma, as well as, total viral nucleic acid loads (viral DNA and RNA) and presence of gross and histological lesions in tissues of various organs collected at the end of the study period (SP).

2. MATERIALS AND METHODS

2.1. Experimental animals and design

Ten mules, arbitrarily identified as they were enrolled from 1 to 10, were acquired from five distinct EIA outbreaks that had occurred in five neighbouring provinces of Central Italy. The age, sex and serological status for EIAV, of the experimental animals recorded on recruitment, are reported in Table 1. The ten animals were divided into two groups on the basis of their AGIDT reactivity observed at the start of the experiment: five mules had either equivocal or negative reactions (Group N) while the remaining animals presented a clear precipitation identity band (Group P), (Scicluna et al., 2013). A negative control, Mule 11, was included in the study that was however kept isolated from the infected animals, but subjected to the same experimental conditions adopted for the other ten mules. Clinical examination and sampling of the negative control were carried out with the same frequency and using the same methods as those described for the infected animals. To ensure its persistent negative status, the EIAV tests on this animal were extended to 30 days after the end of the SP.

The study design was approved by the competent authorities (Italian Ministry of Health – identification No IZS 02/10 RC). Animal husbandry and experimental procedures adopted throughout the SP, that lasted for a minimum of 84 days, were conducted under veterinary supervision and in compliance with the European Union Regulations in force for the use of animals in experiments. Biosecurity measures, as prescribed by the National Regulations, were also adopted to ensure that no EIAV transmission occurred from these animals.

Table 1. Details of the characteristics of the study group: identification, age, gender, serological results observed on recruitment and at the end of OP.

Mule identification no.	Age	Gender	Serological results (ELISA/AGIDT/IB[a])		AGIDT reactivity groups[b]	Duration of IS treatment in days
			On recruitment	At the end of SP		
1	12	F	+/2/+	+/2/+	P	10
2	2	F	+/3/+	+/3/+	P	8
3	30	F	+/0/+	+/0/+	N	8
4	22	F	+/3/+	+/3/+	P	8
5	9	F	+/1/+	+/2/+	N	8
6	8	F	±/0/+	−/0/+	N	8
7	11	C	+/1/+	+/1/+	N	8
8	17	F	+/1/+	+/1/+	N	8
9	11	F	+/2/+	+/4/+	P	8
10	7	F	+/3/+	+/4/+	P	8

[a] Serological results for ELISA and IB are reported qualitatively as positive (+) or negative (−). The AGIDT reactivity is reported as a score as described by Issel et al. (1999).
[b] AGIDT reactivity group: N mules with negative (score 0) or equivocal (score 1) reactivity and P mules with clear reactivity (score 2–5).

2.2. Pharmacologically induced immune suppression

The mules included in the study underwent IS to induce EIAV reactivation. The treatment protocol, described by Kono et al. (1976) and Tumas et al. (1994), started on day 56 from the start of the SP using dexamethasone (Rapison®) at 0.11 mg/kg body weight/die. The duration of the pharmacological administration for each animal was based on the Delayed Type Hypersensitivity (DTH) reaction obtained by the inoculation of *Phaseolus vulgaris*agglutinin® (PHA – Sigma) with the same procedure described by Baus et al. (1996).

2.3. Clinical examination

The general condition of the animals, including the clinical signs considered as characteristic of an acute form of EIA, represented by hyperthermia, diarrhoea, lethargy and anaemia (Leroux et al., 2004), were blindly recorded during a complete physical examination, conducted daily for the whole SP, at the same hour by one clinician to reduce variability in their registration. Threshold for hyperthermia was set at a rectal temperature above 39 °C as reported for horses in similar studies, (Issel et al., 2014).

2.4. Sample collection for laboratory examinations

During the daily clinical examination of the animals, blood, with and without EDTA, were collected from each animal for the laboratory tests described in the present paper. For the estimation of vRNA, plasma was obtained from EDTA blood as follows; volumes of 5 ml of blood were processed on collection by centrifugation at 500×*g* for 5 minutes (min) to separate the plasma from the erythrocytes. The plasma was transferred into a new tube, centrifuged at 1200×*g* for 30 min and the resulting plasma was collected and stored. Nasal swabbing was performed to verify active infections due to equine herpesvirus 1 and 4 and to investigate the possible elimination of EIAV through this route in relation to the presence of the virus in the lungs of affected horses as described by Quinlivan et al. (2007) and Bolfa et al. (2013). On collection, the swabs were immediately immersed in 0.5 ml of phosphate buffer solution (PBS). The supernatant of these samples was collected after centrifugation for 20 min at 1000×*g*. Blood serum was obtained by centrifugation for 10 min at 179×*g*. At the end of the SP, the animals were slaughtered according to the Regulations in force and subjected to necropsy. Tissue samples of the following organs were collected: brain, spinal cord, spleen, lung, heart, mediastinal and mesenteric lymph nodes, liver, kidney and adrenal gland. Samples for virological and serological tests were stored at −80 °C and −20 °C respectively until further examination, and in the case of histological and immunohistochemical analyses, organ samples were processed as described further on.

2.5. Serological and biomolecular methods for pathogens other than EIA

On recruiting, the animals were verified for the presence of infections that could cause a similar clinical picture or influence the evolution of the EIA infection during the study period. The infections considered were based on the local epidemiological situation and clinical pictures similar to EIA that these could induce, as in the case of equine piroplasmosis (EP) caused by *T. equi* and *B. caballi* or immune suppression due to equine herpes virus (EHV) 1 and 4. Testing for presence of these etiological agents was carried out as follows: the serum of the experimental animals was tested for antibodies against equine piroplasms using Babesia equi/caballi Antibody test Kits, cELISA–VMRD, Inc.®. EDTA blood samples were used for the molecular detection of the two protozoa, employing commercial PCR kits, Babesia equi/caballi Genesig® Advanced (PrimerdesignTM Ltd, Southampton, United Kingdom). Animals positive for EP were treated with Imidocarb® (Intervet), a week before the start of the SP, with the dosage indicated for the sterilization of the two protozoa. For EHV 1 and 4, vaccination using Pneumabort K® (Fort Dodge Laboratories) was administered three weeks prior to the start of the SP, followed by a booster four weeks later.

The efficacy of the EP treatment was verified by retesting the subjects using PCR, two weeks after treatment and again weekly throughout the whole SP. Presence of EHV 1 and 4 infections were also monitored on weekly basis, using a Real Time PCR (RT PCR) as described by Damiani et al. (2005).

2.6. Bacteriological examinations

For this purpose, nasal swabs were collected and immediately immersed in Stuart's transport medium and stored at −20 °C for further use. Bacteriological examinations were carried out only if an animal presented clinical signs referable to a respiratory infection of bacterial origin.

2.7. Haematological and PCR methods for EIA

2.7.1. Platelet count (PLT)

The PLT count was performed on EDTA blood, within an hour of its collection, using an automated Cell-Dyn 3700® (Abbott). As no reference baseline for this parameter is available in mules, the value adopted was that reported for horses at 1×10^5 platelets/μl (Issel et al., 2014). All the study animals had pre-SP values above the established baseline during the one-week observation period preceding the start of the experiment (data not shown).

2.7.2. Plasma vRNA loads

The evolution of EIAV replication in the mules was assessed as reported by Cook et al. (2003) and Quinlivan et al. (2007) through the quantitation of the vRNA in the plasma, collected on daily basis. vRNA levels were defined using a qRT PCR as described byScicluna et al., 2013; targeting a region of the exon 1 of the tat gene of EIAV.

Total RNA was extracted in double replicate from a volume of 140 μl of each plasma sample using the QIAamp Viral RNA Mini kit (Qiagen, Hilden, Germany) and employing the automatic extractor QIAcube (Qiagen, Hilden, Germany). Reverse transcription was conducted employing High Capacity cDNA Reverse Transcription Kit (Applied Biosystems, Foster City, CA, USA) on 200 ng of the extracted RNA, concentration established using the NanoDrop 1000 Spectrophotometer (Thermo Fisher Scientific Inc., Wilmington, DE, U.S.A.). The total reaction volume was 60 μl, of which 30 μl contained 200 ng of the extracted RNA, with the addition of the following reagents: 6 μl of 10× random hexamer primers, 6 μl 10X RT-Buffer, 2.4 μl dNTP mix 100 mM, 3 μl of 5 U Multi Scribe Reverse Transcriptase and 12.6 μl of H_2O-DEPC. Synthesis of the cDNA was carried out using AB Veriti 96 well Thermal Cycler (Applied Biosystems, Carlsbad, USA) with the following thermal profile: 25 °C for 10 min, 37 °C for 120 min, 85 °C for 5 min.

The exon 1 of the tat gene of EIAV was used as target of the following primers and probe: MkIII Forward: 5′-GGC GCC CGA ACA GGG ACC-3′. MkIII Reverse: 5′-TGG CCA GGA ACA CCT CCA GAA GAC-3′. The probe employed is a Fluorescent Locked Nucleic Acid (LNA): 5′-FAM-T[+G]A ACC T[+G]G [+C]TG ATC G[+T]A G[+G]A-3′ BHQ. The qRT-PCR, was carried out using the TaqMan Universal PCR Master Mix kit (Applied Biosystems, Foster City, CA, USA) with the following composition: 12.5 μl of TaqMan 2X Universal PCR Master Mix (Applied Biosystems, Foster City, CA, USA), 900 nM of each primer, 300 nM of probe, 5 μl of cDNA and H_2O-DEPC to reach a total volume of 25 μl. The thermal profiles of the qRT-PCR, using AB 7900HT Fast Real-Time PCR System, were as following: 95 °C for 10 min, and 50 cycles of 95 °C for 15 s, 52 °C for 30 s and 60 °C for 1 min. A quantified control, developed as described further on, was included in each qRT-PCR run for the estimation of the vRNA copies/ml of the plasma sample. All data were analysed using the ABI 7900HT Sequence Detection Systems software package ver. 2.4 (Applied Biosystems, Foster City, CA, USA).

The limit of detection (LOD) of the assay was established using the *in vitro* transcribed RNA, obtained from the plasma of Mule 4 using the EIAV MkIII primers. Different Log_{10}concentrations of the transcribed RNA were prepared to cover a range between 3.77×10^9 to 0.2 copies/μl. The limit of quantitation (LOQ) of the qRT-PCR was defined as 3.4 copies/ml while the LOD was 1 copy/ml. The plasma vRNA load for each sample was determined by interpolation of the Ct value on the standard curve and reported as vRNA copies/ml. The Ct value of each plasma sample reported in Fig. 1 represents the mean value obtained from the analyses of the sample extracted and examined as a double replicate.

2.7.3. Total viral nucleic acid in tissues

For the quantitation of the total viral nucleic acid levels, that is vRNA and vDNA in the tissues collected at slaughter, 100 mg of each tissue type were added to 1 ml of PBS and disrupted using FastPrep®

FP120Cell Disrupter Instrument (Qbiogene, Inc. Europe) in Lysing Matrix D tubes (MP Biomedicals, Santa Ana, California, USA). Two hundred µl of the homogenized tissue were employed for nucleic acid extraction with the automatic extractor QIAcube (Qiagen, Hilden, Germany) using the QIAamp cador Pathogen Mini Kit according to the manufacturer instructions. Each sample was extracted in double and the average nucleic acid concentration was measured using a NanoDrop 1000 Spectrophotometer (Thermo Fisher Scientific Inc., Wilmington, DE, U.S.A.). The average number of cells contained in each sample was estimated from the ratio of the mean value of the nucleic acid concentration of the sample and the average cellular content of nucleic acids, using the reference value reported on www.genomesize.com. This was carried out as previously described for the horse by Harrold et al. (2000) and in our case the assumptions used were that cells are diploid and in the Gl phase of the cell cycle (Simmonds et al., 1990). In addition, we assumed that the average cellular content of nucleic acids for the leucocytes of mules, the only data available, is equal to that contained in other tissues. The method used for the quantitation of the total viral nucleic acid loads in tissues was as described for vRNA.

The standardization of the nucleic acid viral loads in tissues was obtained by reporting the quantitative PCR results as the number of viral copies/100,000 cellular equivalent (10^5CE).

2.7.4. Verification of the presence of PCR inhibiting activity

Presence of inhibiting activity in the samples examined by the PCR methods used was verified by the amplification of the endogenous gene coding for Beta-actin, using primers and probe included in the PCR kit of Babesia equi/caballi Genesig® Advanced (PrimerdesignTM Ltd, Southampton, United Kingdom)

2.7.5. Histological and immunohistochemistry procedures

Portions of the organs collected during necropsy were immediately fixed in 10% neutral buffered formalin and submitted to histological examination with the list of examined tissues reported in Table 2. Four-micron tissue sections were stained with hematoxylin-eosin (HE) and the microscopic findings were recorded for the typical EIA-associated lesions (OIE, 2013), hemosiderosis (HS) and lymphomonocytic infiltrates (LI) using the following score system, based on the degree of severity: '–' no lesions; '+' mild (few and scattered lesions); '++' moderate to diffuse (multifocal to disseminated lesions). As for the recording of clinical signs, the same pathologist registered all the histological alterations and had no access to the results of the other tests conducted on the same specimens.

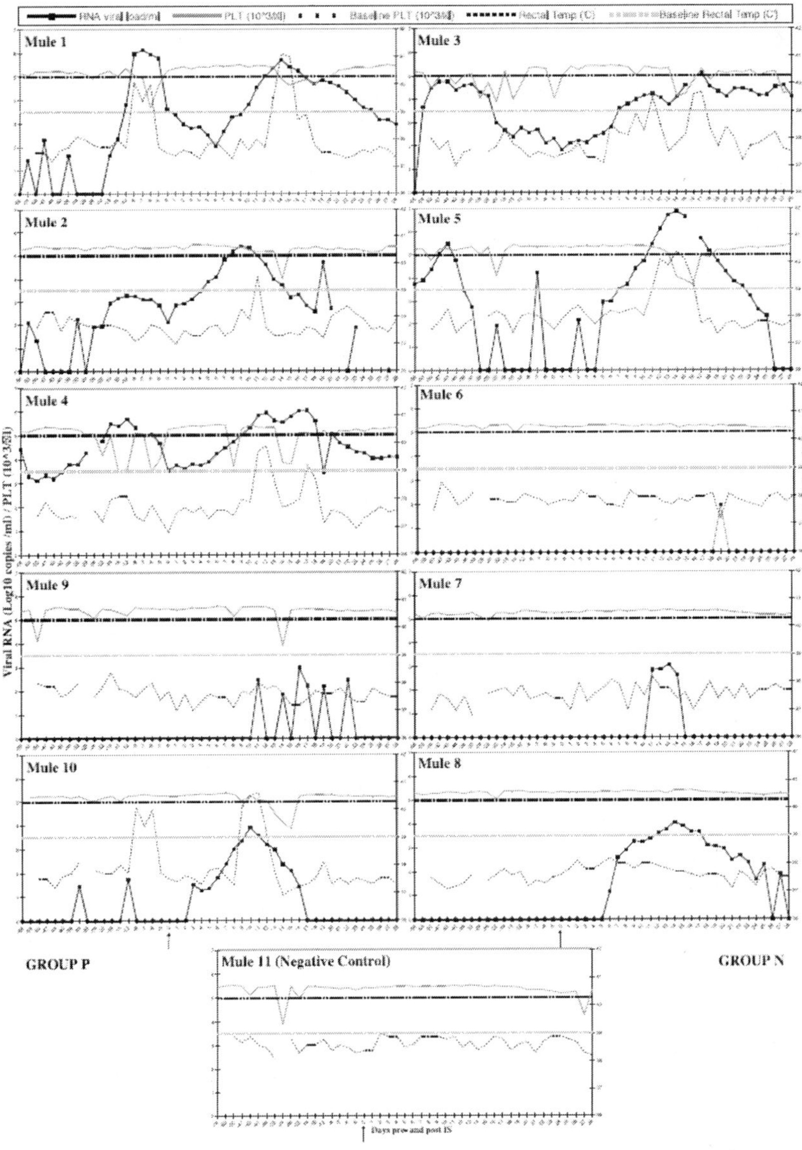

Figure 1. Temporal evolution of body temperature, platelet counts and plasma vRNA load of study animals.

Table 2. Histological, preliminary immunohistochemical (IHC) results and total nucleic acid viral loads in the tissues examined, divided, according to AGIDT reactivity, in Group P (positive), upper half of the table, and Group N (equivocal or negative), lower half.

GROUP P	Mule No 1 (49)				Mule No 2 (70)				Mule No 4 (49)				Mule No 9 (35)				Mule No 10 (35)			
	HISTOLOGY		IHC	Viral Load (copies/ 10^5 CE)	HISTOLOGY		IHC	Viral Load (copies/ 10^5 CE)	HISTOLOGY		IHC	Viral Load (copies/ 10^5 CE)	HISTOLOGY		IHC	Viral Load (copies/ 10^5 CE)	HISTOLOGY		IHC	Viral Load (copies/ 10^5 CE)
	HS	LI			HS	LI			HS	LI			HS	LI			HS	LI		
Adrenal Glands	na	na	na	–	na	na	na	–	–	–	/	1.16	–	–	/	–	–	–	+	1.73
Brain	–	+	/	–	–	–	/	< LOQ	na	na	na	–	–	–	/	–	–	–	/	–
Heart	–	+	/	–	–	–	/	–	–	+	/	0.83	–	–	/	–	–	+	+	–
Kidney	–	+	+	6.95	–	++	/	0.20	–	++	+	–	–	++	/	–	–	+	+	60.83
Liver	+	–	+	5.93	–	–	/	–	–	++	+	14.35	na	na	na	–	–	–	/	3.34
Lung	na	na	na	/	–	+	/	0.47	na	na	na	/	na	na	na	na	na	na	na	/
Mediastinal Lymph Nodes	–	–	/	–	–	–	/	–	–	–	/	1.37	–	–	/	–	–	–	/	–
Mesenteric Lymph Nodes	na	na	na	/	–	–	/	/	na	na	na	/	na	na	na	–	–	–	/	15.54
Spinal Cord	–	–	/	1.79	–	–	/	8.83	na	na	na	11.69	na	na	na	–	–	–	/	–
Spleen	++	–	/	25.19	–	–	/	21.92	++	–	/	0.17	++	–	/	46.86	++	–	–	4.44
GROUP N	**Mule No 3 (28)**				**Mule No 5 (35)**				**Mule No 6 (70)**				**Mule No 7 (70)**				**Mule No 8 (56)**			
	HISTOLOGY		IHC	Viral Load (copies/ 10^5 CE)	HISTOLOGY		IHC	Viral Load (copies/ 10^5 CE)	HISTOLOGY		IHC	Viral Load (copies/ 10^5 CE)	HISTOLOGY		IHC	Viral Load (copies/ 10^5 CE)	HISTOLOGY		IHC	Viral Load (copies/ 10^5 CE)
	HS	LI			HS	LI			HS	LI			HS	LI			HS	LI		
Adrenal Glands	na	na	na	na	–	–	/	4.69	–	+	+	–	na	na	na	–	na	na	na	–
Brain	–	++	?	50.99	na	na	na	0.15	–	+	/	–	–	–	/	–	–	–	/	–
Heart	–	+	/	–	–	+	/	0.08	–	+	/	–	–	+	/	–	–	–	/	–
Kidney	na	na	na	/	–	++	/	0.63	–	+	/	–	–	+	/	–	–	+	/	–
Liver	+	++	+	50.1	–	–	/	29.76	+	–	+	–	–	–	/	0.12	–	–	/	0.22
Lung	–	+	+	–	–	–	+	/	–	–	+	/	–	+	+	0.37	–	–	/	–
Mediastinal Lymph Nodes	+	–	/	6.23	–	–	/	2.51	–	–	–	–	–	–	/	1.3	++	–	/	–
Mesenteric Lymph Nodes	na	na	na	/	–	–	/	9.86	–	–	–	/	–	–	/	/	na	na	na	0.07
Spinal Cord	na	na	na	254.94	na	na	na	–	–	–	/	–	–	–	/	–	–	–	/	–
Spleen	+	–	/	129.25	++	–	/	487	++	–	–	–	+	–	/	40.34	++	–	/	–

Histopathological results are reported with the following score system: '-' no lesions, '+' mild (few and scattered lesions), '++' moderate to diffuse (multifocal to disseminated lesions); HS = hemosiderosis, LI = lymphomonocytic infiltrate. Total nucleic acid viral load are reported as no copies/10^5 CE, <LOQ = inferior to the limit of quantification. '+' positive, '-' negative, '?' presence of aspecific background, 'na' not available, '/' not examined. Number in brackets (n) refers to the day of slaughter after immune suppression.

In a preliminary trial, immunohistochemistry (IHC) was conducted principally on samples with evident histological lesions, using the following procedure: blocking of endogenous peroxidase with 3% H_2O_2 at room temperature for 30 min, antigen retrieving in citrate buffer solution (pH 6) in a water bath at 600 W in a microwave oven for 11 min, followed by an incubation with goat serum (Sigma-Aldrich, Inc., Missouri, US) at room temperature for 1 hour (h), washing in PBS followed by incubation with a primary monoclonal antibody (MAb) anti-p26 (Scicluna et al., 2013) for 1 h at room temperature. A positive reaction was detected using 3,3′-diaminobenzidine (EnVision™ FLEX, Dako, Santa Clara, CA, US) as chromogen with a 3-min development at room temperature and counterstained with haematoxylin. Samples examined and relative results are reported in Table 2 as presence (+) or absence (−) of a positive reaction.

3. RESULTS

As the aim of the study was to evaluate the evolution of clinical, pathological parameters and viral loads of naturally EIA infected mules having equivocal/negative AGIDT reactions, the results of these animals (Group N – Mules 3, 5, 6, 7 and 8) were compared to those presenting a clearly positive AGIDT reactivity (Group P – Mules 1, 2, 4, 9 and 10).

3.1. Immune suppression

Dexamethasone administration was interrupted when the ratio between the DTH skinfold thickness, induced by PHA inoculation with that of the control reaction, was <1 (Kono et al., 1976 and Tumas et al., 1994) indicative of an immune suppressed state. The IS treatment was effective by day 8 in the eleven animals, with the exception of Mule 1, in which IS was obtained on day 10. Side effects due to steroid administration characterised by increased heart rate and laminitis were never observed.

3.2. Clinical, haematological and molecular results

The animals were clinically controlled and sampled on daily basis for 84 days, 56 days before and 28 days after the IS.

Samples that were examined to verify the presence of latent infections for EP and EHV 1 and 4 were confirmed as negative, except for those of Mule 6, that resulted positive for the former in a sample collected on day 12 Post-IS. Despite the spot presence of EHV 1 in this subject, it never presented clinical signs, rise in temperature or thrombocytopenia.

Fig. 1 reports the trends of rectal temperature, PLT and plasma-associated vRNA loads for the evaluation of the temporal association between these alterations relative to each study animal, including that of the negative control.

Hyperthermia, above 39 °C, considered the only clinical sign objectively correlated to the disease (Leroux et al., 2004), was observed between day 8 and 6 Pre-IS, in two mules reaching 40 °C and between day 9 and 17 Post-IS, in six subjects, with a maximum peak of 41.1 °C. In the febrile animals, hyperthermia lasted from one to five days maximum (Fig. 1). Clinical signs were generally absent for the whole observation period with the exception of Mule 3, that frequently presented a profound physical depression throughout the SP.

Following is a description of the alterations observed in the two groups.

Group N: Mule 3 had a constant presence of vRNA for the whole SP with a peak of $\log_{10}5.10$ copies/ml on day 17 Post-IS, concomitant with a febrile episode, and repeated PLT drops below the threshold level, especially in the initial period of the SP and in particular between day 50 Pre- and day 28 Post-IS. In addition to the aforementioned alterations of the sensorium, the subject presented frequent mild anaemia and jaundice with a progressive worsening of its general condition and for this was culled on day 28 Post-IS.

Mule 5 registered a viremic peak of $\log_{10}$ 6.88 copies/ml on day 17 Post-IS that was the highest value observed among all mules, accompanied by fever that was followed by thrombocytopenia lasting for 5 consecutive days, while mild anaemia was mostly present during the second part of the Post-IS period.

In Mules 6, 7 and 8, clinical signs were irrelevant and in the latter two, vRNA loads for the last two animals were detectable only occasionally, at lower levels to those observed in the other mules, between $\log_{10}$ 3.06 and 4.05 copies/ml, while Mule 6 was positive just on day 19 Post-IS, with a low vRNA value of $\log_{10}$ 1.97 copies/ml.

Group P: Mule 1 presented two viremic peaks, $\log_{10}$ 5.69 and 6.13 copies/ml, with the higher observed during the Pre-IS. On both occasions, the viral peaks were accompanied by fever and thrombocytopenia. Only on day 15 Post-IS did the animal present a moderate depression of the sensorium, in concomitance with the second viremic peak, accompanied by a PLT drop that lasted for seven consecutive days.

Mule 2 registered a viremic peak of $\log_{10}$ 5.4 copies/ml, the day following the end of IS, together with a transient low fever and thrombocytopenia, while anaemia was present at the end of the IS.

As for Mule 1, Mule 4 also presented a double vRNA peak, $\log_{10}$ 5.93 and 6.02 copies/ml, with the higher value recorded however during the Post-IS period and had a constant presence of vRNA during the whole SP. Repeated PLT drops in this subject were observed between day 22 Pre-IS and day 15 Post-IS. The same mule presented alterations of the sensorium, with mild depression, anaemia and oedema, subsequently to the second viremic peak.

Mule 9 presented detectable levels of vRNA constantly below $\log_{10}$ 2.96 copies/ml, between day 11 and 22 Post–IS, with just a PLT decline on day 14 Post-IS.

For Mule 10, two febrile episodes were observed; the first during the Pre-SP and the other on day 16 Post-IS, the latter was associated to thrombocytopenia, preceded by a vRNA peak with a value of $\log_{10}$ 3.91 copies/ml.

3.3. Total viral nucleic acid in tissues

The day on which the subjects were slaughtered, as also the results of the total viral nucleic acid/10^5 CE for the different tissues examined, are reported in Table 2. In the majority of the animals examined, the spleen had the highest viral load, followed by the spinal cord, liver, kidney and lymph nodes, with the exception of Mule 3 and 10, in which the highest total nucleic acid values were respectively found in the spinal cord and kidney. The remaining tissues were found positive with values less than 1.6 copies/10^5 CE or undetectable.

In Group N, total viral nucleic acid was also detected in the brain and adrenal glands of two different subjects, Mule 3 and 5, as well as, in the lymph nodes of all animals except for Mule 6. In addition, in this group, the positive animals had overall, higher values than those obtained for the other group, especially for the spleen tissue of Mule 5, with 487 copies/10^5CE. For Group P, the maximum values were detected in the spleen of Mule 10, with 60.83 copies/10^5 CE.

Results of vRNA in nasal swabs were as following: in Mules 1, 2, 4, 5, 6 and 10, detection was sporadic (one to three samples) with vRNA loads close to the detection limit, ranging from 0.002 and 10.74 copies/ml which were recorded between day 1 and 21 Post-IS and were not simultaneous to the plasma vRNA peaks.

3.4. Verification of the presence of PCR inhibiting activity

Amplification of the endogenous gene coding for Beta-actin was successful in all the samples that were examined by the different PCRs described above.

3.5. Bacteriological examinations

No bacterial examinations were carried out on the nasals swabs as there were no signs referable to respiratory diseases during the entire SP.

3.6. Gross, histological and immunohistochemical results

No relevant gross lesions were found at necropsy. The principal microscopic lesions observed are reported in Table 2. Moderate HS in the cytoplasm of macrophages was evident in the spleen of all animals, except for Mule 2. With a lesser degree of severity and at a lower rate, the same alteration was observed in the Kupffer cells and in the lymph nodes macrophages. In the liver of two subjects, lymphomonocyte infiltrates (LI) were present in the periportal duct areas. In the kidney of all subjects examined, LI were present with evidence of mild to moderate membranous proliferative glomerulonephritis in four subjects.

LI was almost constantly present in the lungs examined and occasionally associated with the accumulation of lymphocytes at the peribronchiolar area together with a mild thickening of the alveolar septa. Similar inflammatory lesions of modest entity were also evident in the cardiac interstitium of seven mules. LI were also present in the brain of three animals, Mule 1, 3 and 6, prevalently organised as lymphocytic perivascular cuffings in the white matter (lymphomonocytic leucoencephalitis): in Mule 3, LI were more diffuse with respect to the other two subjects and were also associated with a severe multifocal lymphocytic meningitis. Other lesions observed were the presence of scattered foci of lymphocytic nodules in the adrenal gland of one subject of the five examined and a multifocal or diffuse degeneration of the hepatocytes of Mule 2 and 8.

Preliminary IHC, in some of the tissues examined belonging to three subjects of Group P and four subjects of Group N, detected p26 antigen in the cytoplasm of a few scattered macrophages in the inflammatory lesions of the kidney, liver, lung, heart and adrenal gland (Fig. 2).

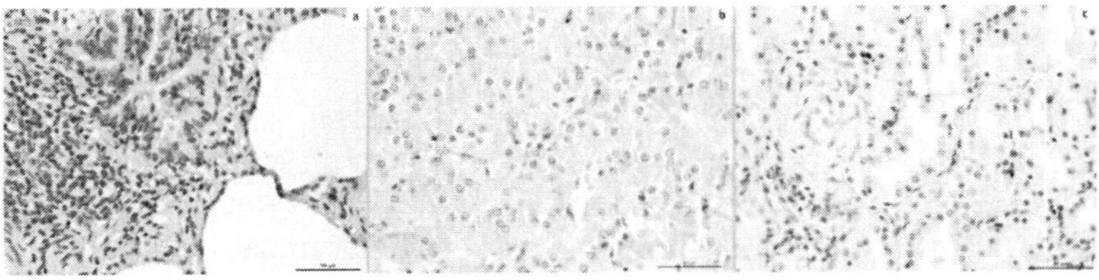

Figure 2. IHC positive reaction: (a) Lung (Mule 5), (b) Adrenal gland and (c) Kidney (Mule 10). Scattered cytoplasmic immunoreactivity in mononuclear cells morphologically resembling macrophages, randomly distributed in (a) the alveolar septa, in (b) the pars intermedia of the adrenal gland and in (c) the renal interstitium. In the lung (a) and in the kidney (c) the immunoreactivity is detectable within the lymphoplasmacytic infiltrates. Anti-p26 Mab, Hematoxylin counterstain, 40×.

The negative control animal did demonstrate any alterations for the clinical, haematological and histological parameters and was constantly virologically and serologically negative for EIAV.

4. DISCUSSION

The current study reports the first description relative to the evolution of the clinical, quantitative virological and pathological findings in ten asymptomatic, naturally infected mules, presenting different AGIDT serological reactivity patterns that were subjected to pharmacological immune suppression for the reactivation of the EIA infection. On enrolment, none of the mules were presenting clinical signs referable to EIA, nor were they in the viremic phase of the infection as all of them were reacting negatively for the presence of plasma vRNA. The animals were however confirmed as infected by EIAV, as proviral cDNA was detected in their buffy coat, using a conventional PCR as described byCappelli et al. (2011) and, as mentioned further on, products of this PCR were used for a preliminary sequencing of the strains involved. This condition is ascribable to an inapparent state of infection. In further support of this, EIA serological diagnosis had been carried out at least six months prior to the beginning of the SP and even then, the study animals were in the inapparent phase of the disease. This status is similar to that described by Hammond et al. (2000) for EIA persistently infected horses during the asymptomatic phase, in the absence of stress related conditions (Harrold et al., 2000).

As in other studies conducted for the investigation of EIAV replication, through the quantitation of the vRNA loads in plasma from which the cellular part was removed, the PCR could have also amplified proviral DNA that could have been present in the sample. In our case we did not include treatment with DNAase, as also omitted by other authors (Quinlivan et al., 2007), as in trails we conducted there was a decrease in the amount of the vRNA amplified. Even in the absence of DNAase treatment, the amount of proviral DNA is minimal because, as stated in the introduction, EIAV replication occurs only in differentiated tissue macrophages. In addition, also as reported by Harrold et al. (2000), during the different disease states, peripheral blood mononuclear cells were found to contain less than 1% of the cellular viral burden.

During the present study, active viral replication began in five mules belonging to both AGIDT serological groups (Fig. 1) during the Pre-IS period and coincided to when the subjects were put together. Such an event presumably represented a stress due to the new management these were experiencing. All the animals were subsequently subjected to an experimentally controlled immune suppression, using the same protocol employed in horses (Kono et al., 1976, Tumas et al., 1994 and Craigo et al., 2007), that provoked vRNA peaking in all of them, even if in one subject, virus replication was detected just on one day. Administration of dexamethasone in these animals proved to be effective in inducing immune suppression as verified by their delayed hypersensitivity reaction. This condition caused the recrudescence of EIA simultaneously to IS, with reactivation or increase of EIAV replication, that reached levels higher than those observed in the Pre-IS period, accompanied by fever and PLT decline in nine of ten study subjects mules. Differently from what was observed in similar experiments conducted in horses (Craigo et al., 2007, Tumas et al., 1994 and Kono et al., 1976), in our study, viral peaking was exclusively accompanied by mild signs that only a careful daily clinical examination was capable of detecting. With the exception of Mule 3, which was continually depressed, neither the high temperature, that was however transient, nor the thrombocytopenia recorded in the other nine mules profoundly altered their general state. If a pathogenic threshold was to be determined in the present study, as reported for the horse, this is between $\log_{10}$ 4.20 and 6.88 copies/ml of plasma vRNA, which is inferior to that reported by Cook et al. (2003). Such lower levels could be due to the use of a different PCR method with a lower recovery rate to that reported by the same authors (Cook et al., 2002).

Although the clinical findings observed throughout the SP corroborated the inapparent phase of infection, they did not correlate with the dynamics of EIAV replication, typical of its reactivation.

On the whole, our mules presented a clinical behaviour similar to that described in donkeys which have plasma vRNA burdens that are 1000-fold less than that of the horse (Cook et al., 2001), but with vRNA peaks that can be described as closer to those reported in acute forms or reactivation of EIA in the latter species.

vRNA replication behaviour observed in the present study is another important feature in the evolution of EIA infection in these animals, as there were important fluctuations in the plasma viral levels, with some animals having periods of vRNA negativity, during which testing of the plasma by molecular methods could have falsely led in classifying them as negative.

When comparing the vRNA replication levels throughout the study period in the two groups, even if the proportion of positive samples were greater for Group P, the interval of the plasma vRNA values observed was however similar; within a range of $\log_{10}$ 2.0–6.8 copies/ml for Group N and $\log_{10}$ 3.0–6.1 copies/ml for Group P, with the highest titre observed in the former. On the basis of these observations, it would seem that viral replication is independent of the serological reactivity, at least for p26, which is the antigen used in the AGIDT, as EIAV replication was indifferently observed within the animals belonging to the different groups. This result is in partial contrast with the assumption that the EIA immune status depends on its antigenic stimulation (Cook et al., 2003). In our case, there were mules, Mule 3, 5 and 8, that even if experiencing a continuous EIAV replication, showed an AGIDT antibody response continually low and not attributable to what could be an initial phase of the infection (Table 1).

For this reason, mules having an equivocal serological response and with fluctuating vRNA plasma loads could pass undetected when using tests of low diagnostic sensitivity, as is the case of the AGIDT (Issel et al., 2013) and also in some cases the PCR when virus levels are below the detection limit and/or due to the intrinsic genetic variability of EIAV.

Relative to the quantitation of the total viral nucleic acid loads, we adopted a method that even if different from that described by other authors in horses (Harrold et al., 2000) and in mules (Spyrou et al., 2003), gave overall results generally in agreement with those reported by these authors, with the exception of the nervous tissue, for which discordant results for the horse exist (Harrold et al., 2000, Kim and Casey, 1992 and Kono et al., 1971). Even in the mule, as also reported by Quinlivan et al. (2007) and Harrold et al. (2000) for the horse, the highest viral nucleic acid loads were constantly detected in the spleen tissue of all subjects examined, with the exception of Mule 3 and 10.

Of note is that the greatest tissue viral nucleic acid loads were observed in two animals of Group N, Mule 3 and 5, that are in agreement with their highest vRNA plasma levels. Of particular interest is whether the viral amounts detected in the brain and spinal cord tissue of Mule 3, as well as the total viral loads and histological lesions at this level, are correlated with the severe depression of the sensorium the subject presented throughout the study and was for this the first to be slaughtered.

In relation to the persistence of EIAV in tissues, moderate to high levels of viral nucleic acids were still detectable even in Mule 2 and 4, slaughtered between 30 and 60 days from when they had experienced their highest vRNA peaks. A further confirmation of this is that animals in which viral nucleic acid was absent corresponded to those with relatively lower levels and proportion of positive samples for plasma vRNA, even if this could be due to a low sensitivity and/or specificity of the PCR method employed for the strains involved that should be further analysed. In fact for Mule 6, the animal with the lowest virological/serological reactivity, these results could be attributed to a particular strain of EIAV or to the capability of the animal of efficiently controlling the infection. This hypothesis may be supported by the presence of the p26 antigen observed with IHC in the lung, adrenal gland and liver of the same animal.

A preliminary study that we conducted on the sequencing of the gag gene of the EIAV strains of the study animals, using the method reported by Cappelli et al. (2011), Mule 6 had a sequence identity of 81% with EIAV Rom-4 (GU060662.1) while the other strains had a sequence identity that varied from 77 to 96% to previously identified European EIAV strains. Further phylogeny studies are being carried out on the EIAV strains involved in this study and will be discussed in another article.

Persistence of EIAV for long intervals at tissues levels, as well as in the plasma, implies that notwithstanding the inapparent state of infection and a limited AGIDT reactivity, animals may act as efficient reservoirs in which viral replication reaching high levels can be retriggered under particular conditions such as those described in this paper.

The microscopic lesions observed in the various tissues are similar to those reported in the chronic forms of EIA in ponies (Oaks et al., 2004) and mules (Spyrou et al., 2003) as also the IHC positivity to what was described by Bolfa et al. (2013) for the horse, using immunofluorescence. In relation to this, lesions and IHC reactivity described in the lung require further investigations to verify the possible viral release through the respiratory tract, a transmission route not previously considered for EIA. Even if our findings do not support this assumption, as low viral loads were detected in the nasals swabs of our animals, this is in contrast to what was described by Quinlivan et al. (2007) who reported higher amounts of RNA copies/ml in nasal swab material in symptomatic horses. Detection of different levels of vRNA in this sample type could be due to the species involved or time since infection, which in the case of the Irish outbreak, was very recent, or also, as previously mentioned, due to the sensitivity of the PCR method, employed.

Further to what was reported by Spyrou et al., 2003; this paper describes the presence of inflammatory lesions in the myocardium and brain of some of the animals examined. Pathological alterations in the brain tissue defined as granulomatous ependymitis (McIlwraith and Kitchen, 1978) or as periventricular lymphohistiocytic leukoencephalitis (Oaks et al., 2004) were occasionally described for horses and ponies but not considered as relevant in an EIA infection. Preliminary trials using IHC confirmed that the histological lesions were EIAV-associated except for the alterations observed in the nervous tissue that lack confirmation due to the presence of an aspecific background. Relative to the absence or low amounts of viral nucleic acid in tissues that presented histological lesions and/or IHC positivity, this could be attributed to the degradation that the former could undergo during sample storage or to the sensitivity of the PCR method, in terms of detection limit and the possible genetic variability of the PCR target. On the other hand, the IHC method has as target, an antigen that is less degradable than genetic material and is also highly conserved in EIAV, as only one serotype is currently known for this virus.

In this study, it was not possible to correlate parameters such as age and breed with the evolution of the EIA infection as the set of animals within the groups were heterogeneous. Relative to the sex of the animals, in the mule population, it is the females that are kept as working animals as they are more docile and therefore our study group is representative of the field population structure. In addition, the identification number was arbitrarily assigned according to the succession of the recruitment time; therefore the fact that Mules 1–5 had a more intensive evolution of EIA is casual. As mentioned earlier, even the viral strains infecting these subjects were to some extent genetically different. The evolution of EIA in our study is on individual basis and is a result of the multiple host and viral factors that are in this context difficult to identify but which however does not correlate with the type of AGIDT reactivity.

According to the current European Union Directive (2009/156) on the movement of equidae between member states it is sufficient to declare freedom of disease just on the absence of clinical signs. In the light of the inapparent phase of EIA infection described in horses and mules, such an affirmation does not fully guarantee the true health status of the animal, which also might not be accurately defined with the sole use of AGIDT. For this, the perception that would rise from the use of a passive surveillance is that in the absence of clinical signs, or a negative AGIDT reaction, an EIA infected equid represents a low transmission risk.

The increasing knowledge on the evolution of EIA, even in mules, as also the data generated from the surveillance conducted in Italy over a period of six years (Scicluna et al., 2013) provided the basis for the set up of an active risk based serological surveillance, using the ELISA as a screening test.

COMPETING INTERESTS

The authors declare that they have no competing interests.

CHAPTER 2

Sampling Strategies in Antimicrobial Resistance Monitoring: Evaluating How Precision and Sensitivity Vary with the Number of Animals Sampled per Farm

Takehisa Yamamoto[1] *, Yoko Hayama[1] , Arata Hidano[1] , Sota Kobayashi[1] , Norihiko Muroga[1], Kiyoyasu Ishikawa[2] , Aki Ogura[2] , Toshiyuki Tsutsui[1]

1 Viral Disease and Epidemiology Research Division, National Institute of Animal Health, Kannondai, Tsukuba, Ibaraki, Japan,

2 Animal Products Safety Division, Food Safety and Consumer Affairs Bureau, Ministry of Agriculture, Forestry and Fisheries, Chiyoda-ku, Tokyo Japan

ABSTRACT

Because antimicrobial resistance in food-producing animals is a major public health concern, many countries have implemented antimicrobial monitoring systems at a national level. When designing a sampling scheme for antimicrobial resistance monitoring, it is necessary to consider both cost effectiveness and statistical plausibility. In this study, we examined how sampling scheme precision and sensitivity can vary with the number of animals sampled from each farm, while keeping the overall sample size constant to avoid additional sampling costs. Five sampling strategies were investigated. These employed 1, 2, 3, 4 or 6 animal samples per farm, with a total of 12 animals sampled in each strategy. A total of 1,500 *Escherichia coli* isolates from 300 fattening pigs on 30 farms were tested for resistance against 12 antimicrobials. The performance of each sampling strategy was evaluated by bootstrap resampling from the observational data. In the bootstrapping procedure, farms, animals, and isolates were selected randomly with replacement, and a total of 10,000 replications were conducted. For each antimicrobial, we observed that the standard deviation and 2.5–97.5 percentile interval of resistance prevalence were smallest in the sampling strategy that employed 1 animal per farm. The proportion of bootstrap samples that included at least 1 isolate with resistance was also evaluated as an indicator of the sensitivity of the sampling strategy to previously unidentified antimicrobial resistance. The proportion was greatest with 1 sample per farm and decreased with larger samples per farm. We concluded that when the total number of samples is pre-specified, the most precise and sensitive sampling strategy involves collecting 1 sample per farm.

1. INTRODUCTION

The emergence of antimicrobial resistance in animals and animal products is a major concern in human health. Consequently, many countries have implemented national monitoring programs for antimicrobial resistance in food production animals or food products. These monitoring programs are aimed at detecting changes in the trends of resistance prevalence and the emergence of resistant microbes [1]. For the same reasons, the Japanese Veterinary Antimicrobial Resistance Monitoring (JVARM) program, a national program for monitoring antimicrobial resistance, has been conducted since 1999 [2] under the direction of the Ministry of Agriculture, Forestry and Fisheries of Japan (MAFF). As a part of monitoring antimicrobial resistance in bacteria obtained from food-producing healthy animals, the prevalence of *Campylobacter spp.*, *Enterococcus spp.*, and *Escherichia coli* is examined by JVARM. JVARM samples a predetermined number of farms in each prefecture every 2 years. In each prefecture, 6 beef cattle farms, 2 pig-fattening farms, 4 layer farms, and 4 broiler farms are sampled. In each selected farm, a sample is taken from an animal and tested for antimicrobial resistance [3].

When designing and planning antimicrobial monitoring systems, the budgetary and human resources required for sampling need to be minimized. The total cost of sampling depends on staff size, their travelling expenditures, and on-site sample collection procedures. It appears that these factors largely depend on the number of farms that are tested, rather than the number of animals sampled per farm. Generally speaking, collecting multiple samples from a single farm is simpler than collecting samples from multiple farms. This provides a motivation for reducing the number of farms that are sampled by JVARM and, in compensation, increasing the number of samples taken from each farm, such that the total sample size remains unchanged.

In many nations such as the Netherlands [4], Denmark [5], Sweden [6], and Canada, the number of samples collected for testing antimicrobial resistance is limited to 1 per farm. The "single sample per farm" strategy is also recommended by the European Food Safety Authority (EFSA), which states that isolates from the same farm are expected to show a similar pattern of resistance; however, the EFSA cites no specific reference for this statement. In addition, although a number of studies have evaluated sampling strategies to understand within-farm prevalence [7], [8], [9], [10], only a few studies have evaluated sampling strategies to assess resistance prevalence at the inter-farm or national levels resistance prevalence.

We sought to provide evidence that would either support this choice of strategy or suggest a better alternative. More specifically, the aim of this study was to determine, on the basis of empirical data, how the number of samples per farm affected the precision and sensitivity of sampling strategies for antimicrobial resistance testing.

2. MATERIALS AND METHODS

2.1 Overview

A total of 1,500 *E. coli* isolates from 30 pig farms in Japan were tested for antimicrobial resistance against 12 drugs. These observational data were used for a simulation study of different sampling strategies. Specifically, we simulated 5 sampling strategies using a bootstrap procedure. Each strategy included a different number of samples per farm, but the same total number of animals sampled (12 animals). The precision of each strategy was evaluated by the deviation of the estimated resistance prevalence. Additionally, the sensitivity of each strategy (the probability of detecting the emergence of antimicrobial resistance that has not been identified previously) was evaluated by the probability that at least 1 resistant isolate was detected.

2.2 Fecal sampling and antibiotic susceptibility testing

Thirty farrow-to-finish pig farms located in 4 prefectures in Japan were selected and sampled between September and October 2011. In each farm, fecal samples were collected from 10 randomly selected fattening pigs aged 150–180 days. Sampling was conducted by field veterinarians with sufficient attention to avoid unnecessary harm to animals. In addition, sampling was conducted under the authorization from farm owners and with the consent of MAFF. Fecal samples were plated on ChromoCult® Coliform Agar (Merck, Darmstadt, Germany), and candidate colonies were confirmed by IMViC (indole, methyl red, Voges-Proskauer, and citrate) reactions. Five *E. coli* isolates were obtained from each fecal sample. All E. coli isolates were tested for antimicrobial resistance using a disc diffusion method [11] with 12 antibiotic disks: oxytetracycline (30 μg; Becton-Dickinson, Franklin Lakes, NJ), sulfisoxazole (250 μg; Becton-Dickinson), sulfamethoxazole plus trimethoprim (23.75 μg plus 1.25 μg, respectively; Becton-Dickinson), ampicillin (10 μg; Nissui Pharmaceutical, Tokyo, Japan), chloramphenicol (30 μg; Nissui), dihydrostreptomycin (10 μg; Nissui), kanamycin (30 μg; Nissui), nalidixic acid (30 μg; Nissui), cefazolin (30 μg; Nissui), gentamicin (10 μg; Nissui), ciprofloxacin (5 μg; Nissui), and cefotaxime (30 μg; Nissui). Antibiotic resistance was determined by the threshold diameter of resistant circles, as described in the instructions attached to the disks. The threshold diameters for each of the antimicrobials were 19, 17, 16, 17, 18, 15, 18, 19, 18, 15, 21 and 23 mm, respectively. The intermediate results were included in the susceptible group, and the results were dichotomized as resistant or susceptible to the different antimicrobials.

2.3 Bootstrap procedure

We compared the precision of different sampling strategies, each of which included a total of 12 animals. We chose to evaluate 12 animals as a total sampling size because the observational data were from 30 source farms of *E. coli* isolates and 12 were available under the single sample per farm strategy with a sufficient number of divisors. Additionally, we set the maximum number of samples per farm at 6 because 10 pigs were sampled per farm in the observational data, and the maximum divisor of 12 is 6. Consequently, 5 strategies were evaluated, sampling 1, 2, 3, 4 or 6 animals per farm, and for all strategies, 12 animals in total were examined (Table 1). Because this study does not focus on the number of isolates per animal, we set the number of isolates to 1 per animal for each strategy.

Table 1. List of sampling strategies evaluated by bootstrap resampling.

ID	# Farms	# Animals	# Total
1	12	1	12
2	6	2	12
3	4	3	12
4	3	4	12
5	2	6	12

doi:10.1371/journal.pone.0087147.t001

To evaluate the precision and sensitivity of sampling strategies, we applied a bootstrap procedure, resampling from the observed data uniformly with replacement [12]. To fully simulate the sampling strategies, bootstrap resampling was performed in 3 steps: at the level of farms, animals, and isolates. In each replication, n_{farm} farms were selected from 30 farms, and then n_{animal} animals were selected in each n_{farm} farm, keeping the total number of selected animals at 12. Finally, 1 isolate was selected within each animal, because the number of isolates from an animal was set to be 1, regardless of the strategy. As a result, 12 samples were selected in each bootstrap replication and this procedure was repeated 10,000 times.

The resistance prevalence—the number of resistant isolates divided by the total number of samples (12)—was calculated for each bootstrap sample for each antimicrobial. The median, 2.5th percentile, 97.5th percentile, and standard deviation (SD) of resistance prevalence obtained from 10,000 bootstrap samples were then calculated to evaluate the precision of each sampling strategy. As an indicator of sampling sensitivity, we calculated the proportion of all bootstrap samples (10,000) that included at least 1 resistant isolate. This calculation was done for each antimicrobial drug separately.

All simulations and statistical analysis were performed using R, version 3.0.0. [13].

3. RESULTS

Numbers within parenthesis in the top row of Table 2 present the proportions of *E. coli*isolates resistant to each antimicrobial in the observed data. Resistance prevalence ranged widely—from 0.76 for oxytetracycline to 0.005 for cefotaxime.

Table 2 presents medians, 2.5th percentiles, 97.5th percentiles, and SDs of prevalence of *E. coli* isolates resistant to each antimicrobial obtained, for each of the 5 sampling strategies. The median values were constant between all sampling strategies and are almost the same as the overall resistance prevalence shown in the top row of Table 2. The 2.5–97.5 percentile interval and the SD were smallest in the sampling strategy that employed 1 animal per farm (ID=1) in all antimicrobial resistance tests. For most antimicrobials, both the 2.5–97.5 percentile interval and the SD increased with the number of animals sampled per farm. However, no trend of SDs could be reliably detected for antimicrobials that had a low prevalence, such as cefazolin, gentamicin, ciprofloxacin, and cefotaxime.

Table 3 presents the proportions of bootstrap samples detecting 1 or more *E. coli* isolates by antimicrobial drug and sampling strategy. These proportions were greatest in the single animal per farm sampling strategy. For most antimicrobials, the proportion decreased as the number of animals per farm increased. However, for antimicrobials that had high prevalence of resistance, the bootstrap estimate of the proportion always took the value 1.0, irrespective of the sampling strategy, indicating that at least 1 resistant isolate was obtained in all bootstrap samples.

Table 2. Proportions of *E. coli* isolates resistant to 12 different antimicrobials obtained with 6 different sampling strategies and 10,000 bootstrap replications per strategy.

	OTC (P = 0.761)				SIX (P = 0.635)				ST (P = 0.474)			
ID	Med	2.5th	97.5th	SD	Med	2.5th	97.5th	SD	Med	2.5th	97.5th	SD
1	0.750	0.500	1.000	0.116	0.667	0.417	0.833	0.130	0.500	0.250	0.750	0.131
2	0.750	0.500	1.000	0.129	0.667	0.333	0.917	0.148	0.500	0.167	0.750	0.160
3	0.750	0.500	1.000	0.141	0.667	0.333	0.917	0.164	0.500	0.167	0.833	0.181
4	0.750	0.417	1.000	0.153	0.667	0.250	0.917	0.179	0.500	0.083	0.917	0.205
5	0.750	0.417	1.000	0.172	0.667	0.167	1.000	0.206	0.500	0.000	0.917	0.240
	ABPC (P = 0.425)				CP (P = 0.410)				SM (P = 0.390)			
ID	Med	2.5th	97.5th	SD	Med	2.5th	97.5th	SD	Med	2.5th	97.5th	SD
1	0.417	0.167	0.667	0.135	0.417	0.167	0.667	0.132	0.417	0.167	0.667	0.132
2	0.417	0.167	0.750	0.152	0.417	0.083	0.750	0.154	0.417	0.083	0.667	0.149
3	0.417	0.083	0.750	0.167	0.417	0.083	0.750	0.172	0.417	0.083	0.750	0.162
4	0.417	0.083	0.750	0.179	0.417	0.083	0.833	0.191	0.417	0.083	0.750	0.176
5	0.417	0.083	0.833	0.205	0.417	0.000	0.833	0.221	0.417	0.000	0.750	0.198
	KM (P = 0.133)				NA (P = 0.057)				CEZ (P = 0.017)			
ID	Med	2.5th	97.5th	SD	Med	2.5th	97.5th	SD	Med	2.5th	97.5th	SD
1	0.083	0.000	0.333	0.094	0.083	0.000	0.250	0.067	0.000	0.000	0.083	0.037
2	0.083	0.000	0.333	0.102	0.083	0.000	0.250	0.069	0.000	0.000	0.083	0.037
3	0.083	0.000	0.335	0.107	0.000	0.000	0.250	0.068	0.000	0.000	0.083	0.038
4	0.083	0.000	0.417	0.113	0.000	0.000	0.250	0.071	0.000	0.000	0.083	0.039
5	0.083	0.000	0.417	0.122	0.000	0.000	0.250	0.073	0.000	0.000	0.083	0.039
	GM (P = 0.013)				CPFX (P = 0.009)				CTX (P = 0.007)			
ID	Med	2.5th	97.5th	SD	Med	2.5th	97.5th	SD	Med	2.5th	97.5th	SD
1	0.000	0.000	0.083	0.032	0.000	0.000	0.083	0.027	0.000	0.000	0.083	0.023
2	0.000	0.000	0.083	0.033	0.000	0.000	0.083	0.028	0.000	0.000	0.083	0.024
3	0.000	0.000	0.083	0.034	0.000	0.000	0.083	0.029	0.000	0.000	0.083	0.024
4	0.000	0.000	0.083	0.035	0.000	0.000	0.083	0.032	0.000	0.000	0.083	0.027
5	0.000	0.000	0.083	0.035	0.000	0.000	0.083	0.033	0.000	0.000	0.083	0.027

P = antimicrobial prevalence in 1,500 fecal samples.
ID = identification of sampling strategy shown in Table 1.
OTC = oxytetracycline, SIX = Sulfisoxazole, ST = sulfamethoxazole/trimethoprim, ABPC = ampicillin, CP = chloramphenicol, SM = dihydrostreptomycin, KM = kanamycin, NA = nalidixic acid, CEZ = cefazolin, GM = gentamicin, CPFX = ciprofloxacin, CTX = cefotaxime.

Table 3. Proportion of bootstrap samples detecting one or more *E. coli* isolates resistant to 12 different antimicrobials, obtained with 6 different sampling strategies and 10,000 bootstrap replications per strategy.

ID	OTC	SIX	ST	ABPC	CP	SM	KM	NA	CEZ	GM	CPFX	CTX
1	1.000	1.000	1.000	0.999	0.999	0.998	0.827	0.513	0.190	0.150	0.109	0.080
2	1.000	1.000	0.999	0.998	0.996	0.996	0.806	0.513	0.183	0.148	0.102	0.076
3	1.000	1.000	0.996	0.994	0.991	0.992	0.791	0.498	0.174	0.140	0.098	0.069
4	1.000	1.000	0.989	0.993	0.984	0.988	0.780	0.494	0.175	0.146	0.095	0.070
5	1.000	0.997	0.969	0.981	0.959	0.974	0.739	0.474	0.169	0.133	0.082	0.059

ID = identification of sampling strategy shown in Table 1.
OTC = oxytetracycline, SIX = Sulfisoxazole, ST = sulfamethoxazole/trimethoprim, ABPC = ampicillin, CP = chloramphenicol, SM = dihydrostreptomycin, KM = kanamycin, NA = nalidixic acid, CEZ = cefazolin, GM = gentamicin, CPFX = ciprofloxacin, CTX = cefotaxime.

4. DISCUSSION

One of the major objectives of national antimicrobial resistance monitoring is to describe trends of resistance prevalence in the target animal population [14], [15]. Therefore, this study aimed to evaluate the precision of 5 different sampling strategies based on the deviance of estimated prevalence, as assessed by 2.5th–97.5th percentile intervals and SDs. Statistically speaking, a narrow interval and small SD both suggest that the estimated prevalence is more stable and that similar values will be observed across repeated sampling if the target population is static. Thus, the sampling strategy of 1 sample per farm is the most reliable strategy, offering maximum precision. Precision decreased with the increase in the number of samples taken per farm. In the present study, the data used in our bootstrap procedure were obtained from a sampling scheme that was not entirely random; therefore, the true antimicrobial prevalence in the national population remained unknown. Therefore, precise evaluation of the accuracy of these strategies is difficult. Despite this limitation, the agreement of median prevalence between the 5 strategies may indicate a favorable accuracy.

Another objective of national monitoring is to detect an emergence of antimicrobial resistance that was not previously observed in the population [14], [15]. To evaluate the sensitivity of several sampling strategies, we calculated the proportion of bootstrap samples in which 1 or more resistant isolates were detected. As in our evaluation of precision, the sampling strategy with 1 sample per farm exhibited the best sensitivity, and sensitivity decreased with an increase in the number of samples taken per farm. To our knowledge, this is the only study that evaluates the sensitivity of sampling strategies for antimicrobial resistance monitoring based on empirical data. Our comparison of sampling strategy effectiveness by 2 different indicators—precision and sensitivity—will help convince decision makers and stakeholders of our principal result—the superiority of sampling one animal per farm in monitoring programs.

Our result supports the “single sample per farm” strategy that is currently applied in many countries [4], [5], [6] and recommended by the EFSA. As demonstrated by previous studies targeting farm-level monitoring, the best strategies of resistance prevalence monitoring involve taking 1 sample per animal [16], [8]. As an evaluation of national monitoring, our results agree with those of Regula et al. [17], who also demonstrated the advantage of a “single sample per farm” strategy using a bootstrap sampling procedure. Regula et al. obtained bootstrap samples from an artificial dataset of animal and farm bacterial resistance, which was generated from measurements of resistance prevalence. In contrast, our study employs actual field data of 1,500 *E. coli* isolates. Therefore, we believe that our results reflect the true deviance of resistance at the level of farms, animals, and isolates.

This study is subject to some limitations, and therefore, some caution is warranted when interpreting its results. Farms were not selected at random and the number of sampled animals was limited. Additionally, we did not examine animals other than pigs, nor bacteria other than *E. coli*. Although these factors could affect the generalizability of our results, we believe that these limitations are mitigated by our evaluation of sampling strategies using resistance data for a variety of antimicrobials. Furthermore, the actual cost of sampling should also be considered during discussions of the future sampling strategy for the national monitoring system. Future research could examine the cost-effectiveness of different monitoring program designs and the applicability of our results to different livestock and bacteria.

When designing the sampling strategy for a national monitoring system, researchers should specify the minimum change in resistance prevalence that can be detected by the system, as well as the system's threshold for discovery of newly emergent resistance [14]. When a specific sampling strategy needs to be selected with a fixed total number of samples, the single sample per farm strategy should be recommended.

ACKNOWLEDGMENTS

The authors gratefully acknowledge the Japan Livestock Industry Association (JLIA) for providing the results of antimicrobial susceptibility testing.

AUTHOR CONTRIBUTIONS

Conceived and designed the experiments: TY YH NM TT. Performed the experiments: KI AO. Analyzed the data: TY YH AH SK. Contributed reagents/materials/analysis tools: KI AO. Wrote the paper: TY.

REFERENCES

1. Silley P, Simjee S, Schwarz S (2012). Surveillance and monitoring of antimicrobial resistance and antibiotic consumption in humans and animals. Rev Sci Tech 31: 105–120.

2. Kijima-Tanaka M, Ishihara K, Morioka A, Kojima A, Ohzono T, et al. (2003) A national surveillance of antimicrobial resistance in Escherichia coli isolated from food-producing animals in Japan. J Antimicrob Chemother 51: 447–451.

3. MAFF (Ministry of Agriculture, Forestry and Fisheries, Japan) (2009) A report on the Japanese veterinary antimicrobial resistance monitoring system -2000 to 2007-. Available: http://www. maff. go. jp/nval/tyosa_kenkyu/taiseiki/pdf/ jvarm2000_2007_ final_ 201005. pdf. Accessed 2013 May 10.

4. CVI (Central Veterinary Institute of Wageningen University) (2012) MARAN 2012: Monitoring of antimicrobial resistance and antibiotic usage in animals in the Netherlands in 2010/2011. Available: http://www.wageningenur.nl/web/ file?uuid = 5ced1ff9-1ba2-465d-918f-66fc7774d176&owner = 57e2893d-bcff- 4965-977b-ca5f23f047a5. Accessed 2013 May 10.

5. DANMAP (Danish Integrated Antimicrobial Resistance Monitoring and Research Programme) (2011) DANMAP 2011. DANMAP, Danish integrated antimicrobial resistance monitoring and research programme, Denmark. Available: http://www.danmap.org/Downloads/~/media/Projekt%20sites/ Danmap/DANMAP%20reports/Danmap_2011.ashx. Accessed 2013 Aug 27.

6. SVA (National Veterinary Institute, Sweden) (2011) Swedish veterinary antimicrobial resistance monitoring - SVARM (2011). Available: http://www. sva.se/upload/Redesign2011/Pdf/Om_SVA/ publikationer/Trycksaker/ Svarm2011.pdf. Accessed 2013 May 10.

7. Dunlop RH, McEwen SA, Meek AH, Friendship RM, Black WD, et al. (1999) Sampling considerations for herd-level measurement of faecal Escherichia coli antimicrobial resistance in finisher pigs. Epidemiol Infect 122: 485–496.

8. Persoons D, Bollaerts K, Smet A, Herman L, Heyndrickx M, et al. (2011) The importance of sample size in the determination of a

flock-level antimicrobial resistance profile for Escherichia coli in broilers. Microb Drug Resist 17: 513– 519.

9. Vieira AR, Wu S, Jensen LB, Dalsgaard A, Houe H, et al. (2008) Using data on resistance prevalence per sample in the surveillance of antimicrobial resistance. J Antimicrob Chemother 62: 535–538.

10. Villarroel A, Morley PS, Wittum TE, Bolte DS (2006) Use of a simulation model to evaluate sampling strategies for characterization of antimicrobial resistance in non-type-specific Escherichia coli isolated from dairy cows. Am J Vet Res 67: 951–956.

11. Clinical and Laboratory Standard Institute (2006) Performance standards for antimicrobial susceptibility tests-ninth edition: approved standard M02-A9. Clinical and Laboratory Standards Institute, Wayne, PA.

12. Efron B, Tibshirani RJ (1993) An introduction to the bootstrap. New York: Chapman & Hall.

13. R Core Team (2013) R: A Language and Environment for Statistical Computing. R Foundation for Statistical Computing, Vienna, Austria. Available: http://www.R-project.org/. Accessed 2013 Aug 27.

14. EFSA (European Food Safety Authority) (2012) Technical specifications on the harmonised monitoring and reporting of antimicrobial resistance in Salmonella, Campylobacter and indicator Escherichia coli and Enterococcus spp. Bacteria transmitted through food. Available: http://www.efsa.europa.eu/en/search/ doc/2742.pdf. Accessed 2013 May 10.

15. OIE (World Organization for Animal Health) (2012) Terrestrial Animal Health Code; CHAPTER 6.7. Harmonisation of national antimicrobial resistance surveillance and monitoring programmes. Available: http://www.oie.int/index. php?id = 169&L = 0&htmfile = chapitre_1.6.7.htm. Accessed 2013 May 10.

16. Brun E, Holstad G, Kruse H, Jarp J (2002) Within-sample and between-sample variation of antimicrobial resistance in fecal Escherichia coli isolates from pigs. Microb Drug Resist 8: 385–391.

17. Regula G, Lo FDM, Ledergerber U, Stephan R, Danuser J, et al. (2005) Evaluation of an antimicrobial resistance monitoring program for Campylobacter in poultry by simulation. Prev Vet Med 70: 29–43.

CHAPTER 3

Blood sample collection in small laboratory animals

Parasuraman S, Raveendran R, Kesavan R

Department of the Pharmacology, Jawaharlal Institute of Postgraduate Medical Education and Research, Pondicherry, India

Collection of blood from small laboratory animals is necessary for a wide range of scientific research and there are a number of efficient methods available for that. It is important that blood sample collection from experimental animals should be least stressful because stress will affect the outcome of the study. Various regulatory agencies and guidelines have restricted the use of animals and the techniques used for blood collection in laboratory animals. This article deals with the approved blood collection techniques for laboratory animals like rodents, lagomorphs and nonrodents. Permission of the Institute Animal Ethics Committee has been obtained for the use of animals for demonstrating the techniques.

1. GENERAL PRINCIPLES OF BLOOD COLLECTION IN ANIMALS

- The method of blood collection should be described in the protocol approved by the Institute animal ethics committee.
- It should be least painful and stressful. Blood sample may be collected under anesthesia [Table 1] or without anesthesia.[1]

Table 1 Commonly recommended anesthetic agents for laboratory animal experiments

Animal species	Short anesthesia	Medium anesthesia	Long anesthesia
Mice	Isoflurane (inhalation) Halothane (inhalation)	Xylazine + ketamine (5 mg + 100 mg i.m.)	Xylazine + ketamine (16 mg+60 mg i.m./i.p.)
Rat		Xylazine + ketamine (5 mg + 100 mg i.m.)	Xylazine + ketamine (16 mg +60 mg i.m./ i.p.) or Urethane (1200 mg/kg i.p.)
Guinea pig	Isoflurane (inhalation)	Xylazine + ketamine (2 mg + 80 mg i.m.)	Xylazine + ketamine (4 mg + 100 mg i.m.)
Rabbits	Isoflurane (inhalation)	Xylazine + ketamine (5 mg + 15 – 30 mg i.m.)	Xylazine + ketamine (5 mg + 100 mg i.m.)

Atropine (0.02 mg/kg s.c./i.m.) is used as a preanesthetic medication for all the species to reduce salivation, bronchial secretion and protect heart from vagal inhibition.[6-8]

- Adequate training is required for blood collection using any method in any species.

- In general, blood sample is withdrawn from venous, arterial blood vessels or heart chambers.

- Frequency of blood collection is important. Once in two weeks is ideal for nonrodents. If the study needs multiple blood samples, lagomorphs (e.g., hares and rabbit) can be used.

- All nonterminal blood collection without replacement of fluids is limited up to 10% of total circulating blood volume in healthy, normal, adult animals on a single occasion and collection may be repeated after 3 to 4 weeks. In case repeated blood samples are required at short intervals, a maximum of 0.6 ml/kg/day or 1.0% of an animal's total blood volume can be removed every 24 hour.[2,3]

- If the study involves repeated blood sample collection, the samples can be withdrawn through a temporary cannula. This may reduce pain and stress in the experimental animals.

- The estimated blood volume in adult animals is 55 to 70 ml/kg body weight. Care should be taken for older and obese animals.[4] If blood collection volume exceeds more than 10% of total blood volume, fluid replacement may be required. Lactated Ringer's solution (LRS) is recommended as the best fluid replacement by National Institutes of Health (NIH). If the volume of blood collection exceeds more than 30% of the total circulatory blood volume, adequate care should be taken so that the animal does not suffer from hypovolemia.[5]

2. GENERAL METHODS FOR BLOOD COLLECTION

Blood samples are collected using the following techniques:[1]

- Blood collection not requiring anesthesia
  - Saphenous vein (rat, mice, guinea pig)
  - Dorsal pedal vein (rat, mice)

- Blood collection requiring anesthesia (local/general anesthesia)
    - Tail vein (rat, mice)
    - Tail snip (mice)
    - Orbital sinus (rat, mice)
    - Jugular vein (rat, mice)
    - Temporary cannula (rat, mice)
    - Blood vessel cannulation (rat, guinea pig, ferret)
    - Tarsal vein (guinea pig)
    - Marginal ear vein/artery (rabbit)
- Terminal procedure
    - Cardiac puncture (rat, mice, guinea pig, rabbit, ferret)
    - Orbital sinus (rat, mice)
    - Posterior vena cava (rat, mice)

3. PROCEDURE FOR SAPHENOUS VEIN BLOOD SAMPLE COLLECTION[9]

Requirements include animal, rodent handling gloves, towel, cotton, sample collection tubes and 20G needle.

- Lateral saphenous vein is used for sampling while taking aseptic precautions.
- The back of the hind leg is shaved with electric trimmer until saphenous vein is visible. Hair removal cream can also be used.
- The animal is restrained manually or using a suitable animal restrainer.
- Hind leg is immobilized and slight pressure may be applied gently above the knee joint.

- The vein is punctured using a 20G needle and enough volume of blood is collected with a capillary tube or a syringe with a needle. The punctured site is compressed to stop the bleeding. While collecting blood:
    - the local anesthetic cream may be applied on the collection site
    - no more than three attempts are made
    - continuous sampling should be avoided and
    - collecting more than four samples in a day (24-hour period) is not advisable.

4. PROCEDURE FOR DORSAL PEDAL VEIN BLOOD SAMPLE COLLECTION

Requirements include animal (rat or mice), rodent handling gloves, cotton, capillary tube, 23G/27G needle and blood sample collection tubes.

- The animal is kept in a restrainer.
- The hind foot around ankle is held and medial dorsal pedal vessel is located on top of the foot.
- The foot is cleaned with absolute alcohol and dorsal pedal vein is punctured with 23G/27G needle.
- Drops of blood that would appear on the skin surface are collected in a capillary tube and a little pressure is applied to stop the bleeding [Figure 1].

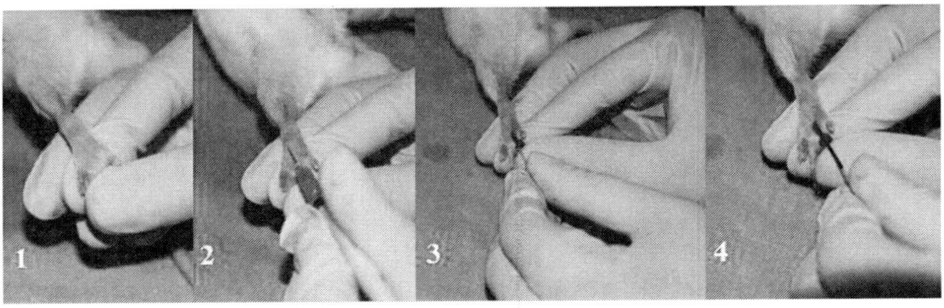

Figure 1. Blood sample collection from rat dorsal pedal vein

5. PROCEDURE FOR TAIL VEIN BLOOD SAMPLE COLLECTION

Requirements include animal, rodent handling gloves, towel, cotton, sample collection tube and animal warming chamber.

- This method is recommended for collecting a large volume of blood sample (up to 2ml/withdrawal)
- The animal is made comfortable in a restrainer while maintaining the temperature around at 24 to 27°C.
- The tail should not be rubbed from the base to the tip as it will result in leukocytosis. If the vein is not visible, the tail is dipped into warm water (40°C).
- Local aesthetic cream must be applied on the surface of the tail 30 min before the experiment.
- A 23G needle is inserted into the blood vessel and blood is collected using a capillary tube or a syringe with a needle. In case of difficulties, 0.5 to 1 cm of surface of the skin is cut open and the vein is pricked with bleeding lancet or needle and blood is collected with a capillary tube or a syringe with a needle.
- Having completed blood collection, pressure/silver nitrate ointment/solution is applied to stop the bleeding.
- If multiple samples are needed, temporary surgical cannula may be used.

- Restrainer is washed frequently to avoid/prevent pheromonally induced stress or cross infection [Figure 2].

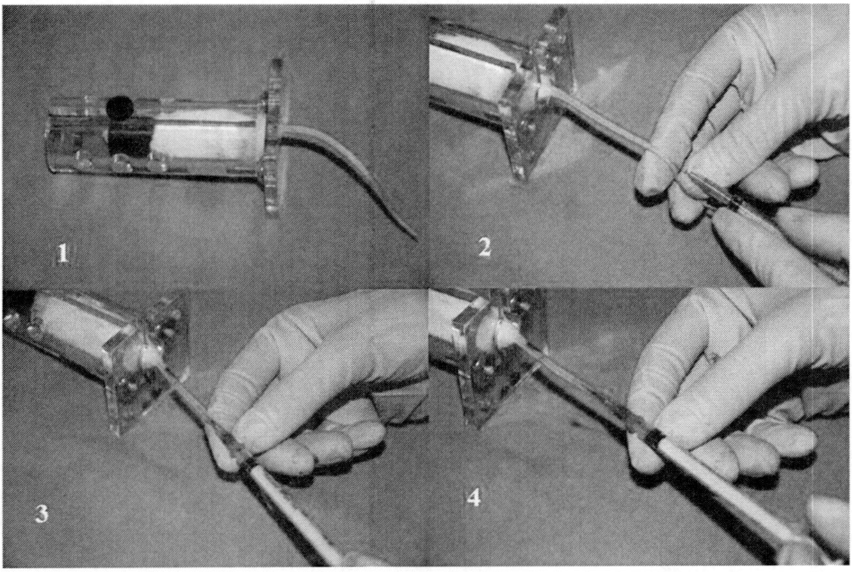

Figure 2. Blood sample collection from mouse tail vein

6. PROCEDURE FOR TAIL SNIP BLOOD SAMPLE COLLECTION

Requirements include animal, anesthetic agent, cotton, surgical blade and blood sample collection tubes.

- This method is recommended for blood collection only in mice.

- This method should be avoided as far as possible because it can cause potential permanent damage on the animal tail. If needed, it should be done under terminal anesthesia only.

- Before collecting the blood, local anesthesia is applied on the tail and a cut is made 1 mm from the tip of the tail using scalpel blade.

- Blood flow is stopped by dabbing the tail tip.

7. PROCEDURE FOR ORBITAL SINUS BLOOD SAMPLE COLLECTION

Requirements include animal, anesthetic agent, cotton, capillary tube and blood sample collection tubes.

- This technique is used with recovery in experimental circumstances and this method is also called periorbital, posterior-orbital and orbital venous plexus bleeding.
- Blood sample is collected under general anesthesia.
- Topical ophthalmic anesthetic agent is applied to the eye before bleeding.
- The animal is scruffed with thumb and forefinger of the nondominant hand and the skin around the eye is pulled taut.
- A capillary is inserted into the medial canthus of the eye (30 degree angle to the nose).
- Slight thumb pressure is enough to puncture the tissue and enter the plexus/sinus.
- Once the plexus/sinus is punctured, blood will come through the capillary tube.
- Once the required volume of blood is collected from plexus, the capillary tube is gently removed and wiped with sterile cotton. Bleeding can be stopped by applying gentle finger pressure.
- Thirty minutes after blood collection, animal is checked for postoperative and periorbital lesions [Figures [Figures33 and and44].
- Caution:
    - Repeated blood sampling is not recommended.
    - Skill is required to collect blood.
    - Even a minor mistake will cause damage to the eyes.
    - Two weeks should be allowed between two bleedings.

- Adverse effects reported from this method is around 1 to 2% which includes hematoma, corneal ulceration, keratitis, pannus formation, rupture of the globe, damage of the optic nerve and other intraorbital structures and necrotic dacryoadenitis of the harderian gland.

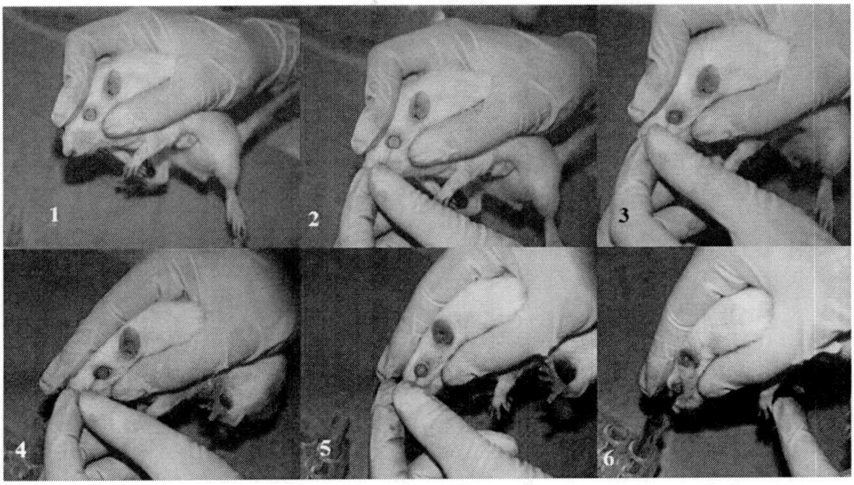

Figure 3. Blood sample collection from rat orbital sinus

8. PROCEDURE FOR JUGULAR VEIN BLOOD SAMPLE COLLECTION

Requirements include animal, anesthetic agent, cotton, 25G needle and blood sample collection tubes.

- In this method, warming of the animals is not required and is used to collect micro volumes to one ml of blood sample.
- This method has to be carried out under general/inhalation anesthesia and two persons are needed to collect blood sample.
- One person has to restrain the animal and monitor the animal. Another person is required to collect the blood sample from the animal.

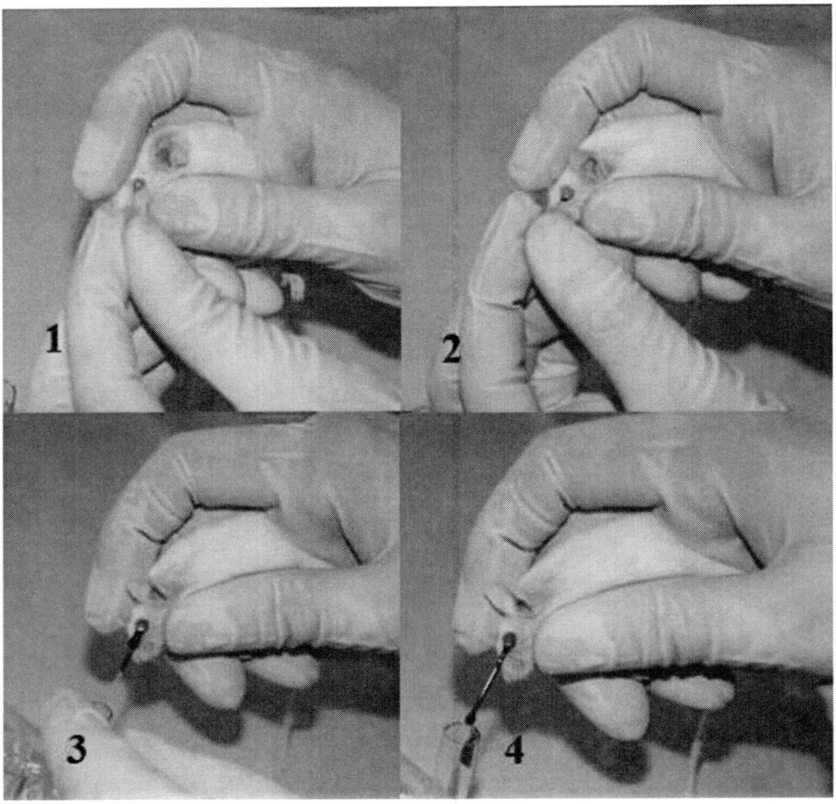

Figure 4. Blood sample collection from mouse orbital sinus

- The neck region of the animal is shaved and kept in hyperextended position. The jugular veins appear blue in color and is found 2 to 4 mm lateral to sternoclavicular junction. A 25G needle is inserted in the caudocephalic direction (back to front) and blood is withdrawn slowly to avoid collapse of these small blood vessels. Animal has to be handled carefully and not more than 3 to 4 mm of needle is to be inserted into the blood vessel.

- If the attempt to collect blood fails, the needle is slowly removed and the site is monitored for bleeding. If there is no bleeding, one more attempt can be made. Further attempts should be avoided in case of bleeding as it may collapse the vein.

- Finger pressure is applied to stop bleeding.

- Caution:
  - Number of attempts is limited to three.
  - Apply local anesthetic cream 30 minutes prior to sampling.

9. PROCEDURE FOR BLOOD SAMPLE COLLECTION WITH TEMPORARY CANNULA

Requirements include animal, anesthetic agent, cotton, 25G needle, animal warming chamber and blood sample collection tubes.

- Usually a temporary cannulation is made in the tail vein and used for a few hours.
- The animal is restrained and local anesthetic cream is applied on the tail (1 – 2 cm above the tail tip).
- The tail is either cannulated or a 25G needle is used.
- Tail bleeding normally requires the animal to be warmed in order to dilate the blood vessels (37 – 39°C for 5 – 15 min).
- After cannulation, animal has to be housed individually in large cages.

10. PROTOCOL FOR BLOOD VESSEL CANNULATION

Requirements include animal, anesthetic agent, cotton, 25G needle, i.v. cannula, surgical blade, heparin (or any anticoagulant) and blood sample collection tubes.

- This method involves continuous and multiple sampling in the experimental animal.
- This method requires close and continuous monitoring of the animal.

- Usually blood vessel cannulation is done in the femoral artery, femoral vein, carotid artery, jugular vein, vena cava and dorsal aorta.
- Surgery is required for this method and appropriate anesthesia and analgesia should be used to minimize the pain.
- After surgical cannulation, animal should be housed singly in a large and spacious cage.
- Blood sample may be collected over 24 hour at the volume of 0.1 to 0.2 ml/sample.
- After withdrawing the blood, the cannula is flushed with an anticoagulant and the withdrawn volume may be replaced (if required) with LRS and cannula should be closed tightly [Figure 5].

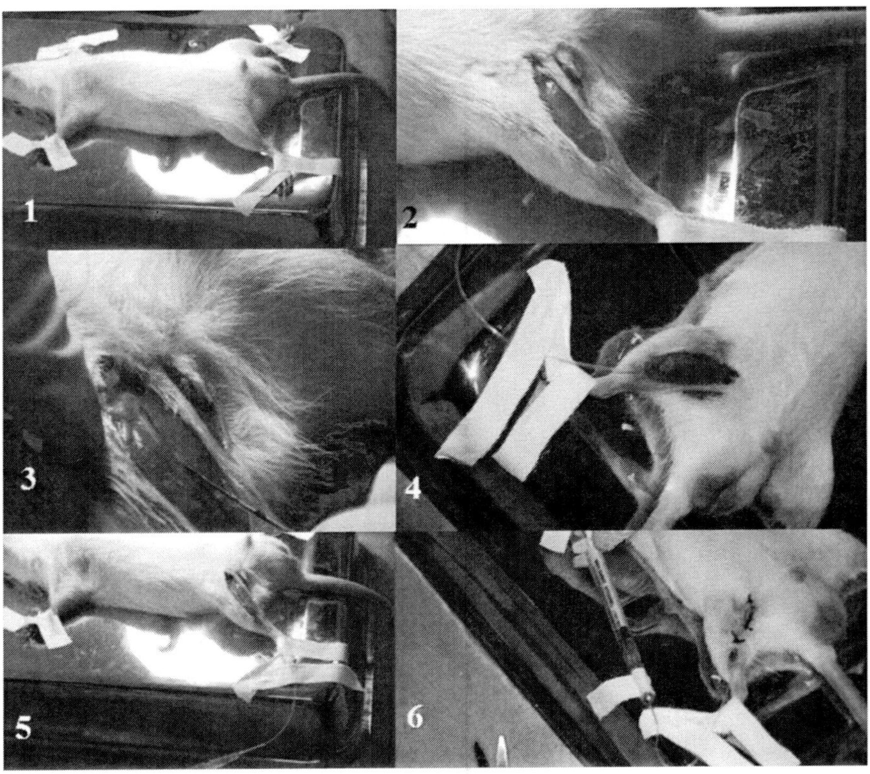

Figure 5. Blood vessel cannulation of rat femoral vein

- Caution: The experiment has to be conducted fully under aseptic precautions. Infection, hemorrhage, blockage of cannula and swelling around the cannulation site should be looked for. The needle size and maximum blood volume to be collected are given in Table 2.

Table 2. Needle size used for blood vessel cannulation in different species

Species	Needle to be used	Maximum collection volume
Mice	23 – 25G	1 ml
Rat	19 – 21G	10 – 15 ml
Rabbit	19 – 21G	60 – 200 ml
Guinea pig	20 – 21G	1 – 25 ml

11. PROTOCOL FOR TARSAL VEIN BLOOD SAMPLE COLLECTION

Requirements include animal, anesthetic agent, cotton, 22G needle, hair remover and blood sample collection tubes.

- Tarsal vein is identified in one of the hind legs of large animals. This method is commonly recommended for guinea pig.
- One person has to restrain the animal properly. Tarsal vein may be visible in blue color.
- The surface hairs are removed by applying a suitable hair remover. A local anesthetic cream is applied on the collection site.
- After 20 to 30 minutes, blood sample is collected slowly by using 22G needle.
- Maximum three samples can be taken per leg and 0.1 to 0.3 ml of blood can be collected per sample.

- After the sample collection, gentle pressure is applied with finger for 2 minutes to stop bleeding [Figure 6].
- Caution:
    - Not more than six samples from both hind legs are taken.
    - The number of attempts is three or less.

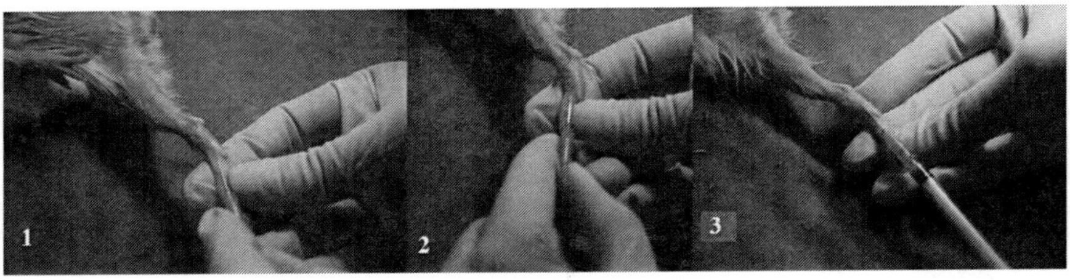

Figure 6. Blood sample collection from guinea pig tarsal vein

12. PROTOCOL FOR MARGINAL EAR VEIN/ARTERY BLOOD SAMPLE COLLECTION

Requirements include animal, anesthetic agent, cotton, 26G needle, 95% v/v alcohol, o-Xylene, surgical blade and blood sample collection tube.

- This method is commonly adopted for rabbits.
- The animal should be placed in a restrainer.
- Ear is cleaned with 95% v/v alcohol and local anesthetic cream is applied on the collection site 10 min prior to sampling. (If required, the o-Xylene/topical vasodilator may be applied topically on the collection site to dilate blood vessels).
- Size 11 surgical blade is used to cut the marginal ear vein and blood is collected in a collecting tube. Otherwise, a 26G needle may be used to collect blood from animal marginal vein.

- After collecting blood, clean sterile cotton is kept on the collection site and finger pressure is applied to stop the bleeding [Figures [Figures77 and and88].

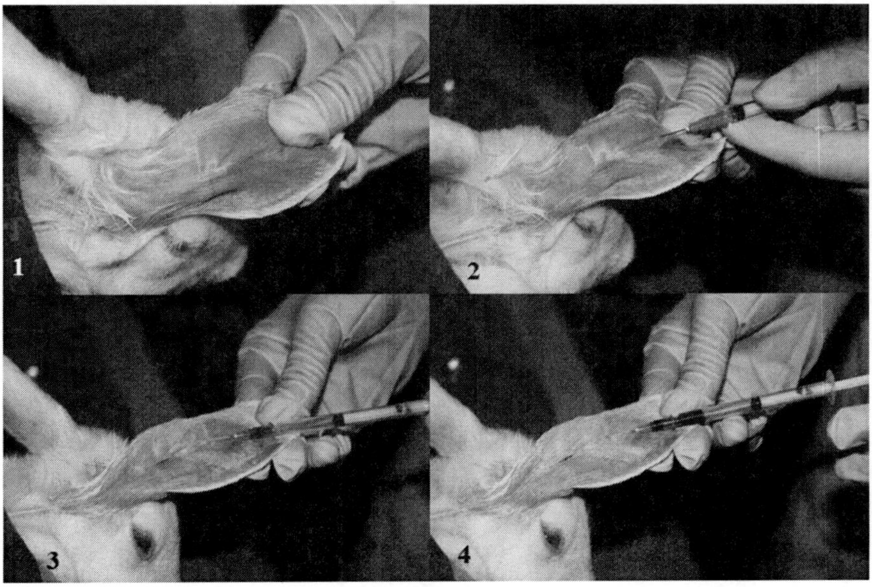

Figure 7. Blood sample collection from rabbit marginal ear vein using 26 G needle

13. PROTOCOL FOR CARDIAC PUNCTURE[10,11]

Requirements include animal, anesthetic agent, towel, cotton, 19 to 25G needle with 1 to 5 ml syringe, surgical blade, tube (internal diameter of 0.1 to 0.3 mm) for thoracotomy, plastic disposable bag and blood sample collection tubes.

- In general, cardiac puncture is recommended for terminal stage of the study to collect a single, good quality and large volume of blood from the experimental animals.

- During blood sample collection, animal will be in terminal anesthesia.

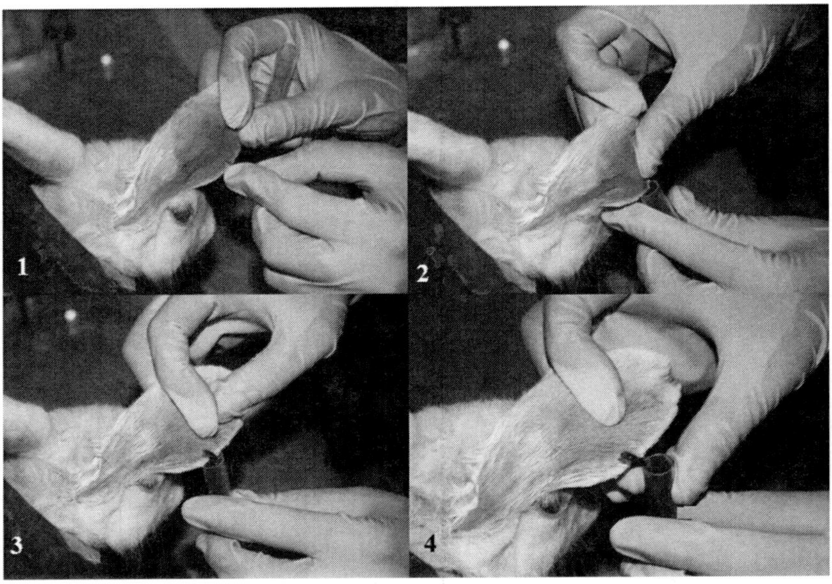

Figure 8. Blood sample collection from rabbit marginal ear vein using incision method.

- Appropriate needle is used for blood sample collection with or without thoracotomy. Blood sample will be taken from the heart, preferably from the ventricle slowly to avoid collapsing of heart [Figure 9].
- Caution: If animal has dextrocardia, sampling may fail.

14. PROTOCOL FOR BLOOD SAMPLE COLLECTION THROUGH POSTERIOR VENA CAVA

Requirements include animal, anesthetic agent, surgical blade, small glass rods, surgical scissor, 21 to 25G needle with 1 to 5 ml syringe and blood sample collection tube.

- In general, posterior vena cava blood sample is recommended for terminal stage of the study.

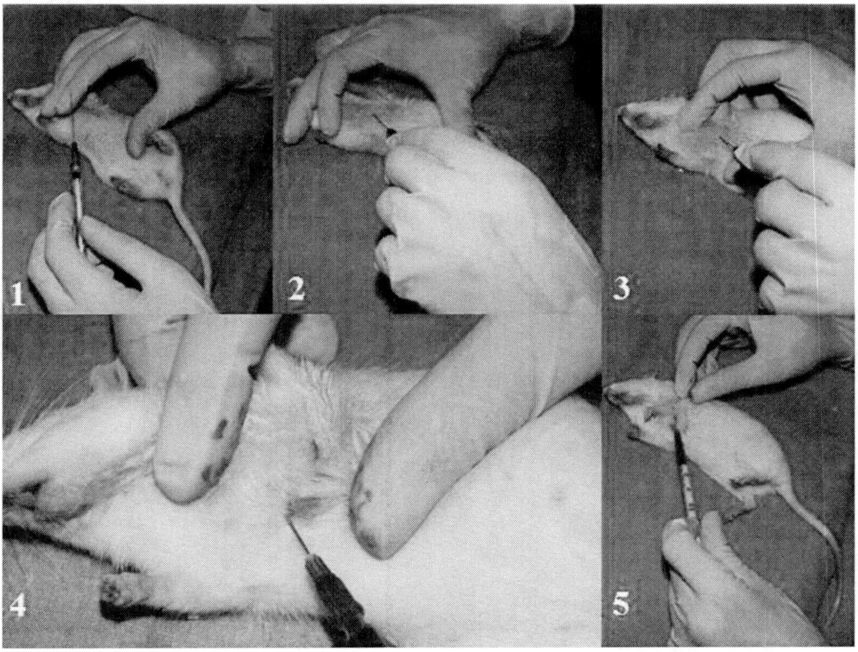

Figure 9. Blood sample collection through cardiac puncture in rat

- Animal have to be anesthetized and 'Y'- or 'V'-shaped cut in the abdomen is made and the intestines are gently removed.
- The liver is pushed forward and the posterior vena cava (between the kidneys) is identified.
- 21 to 25G needle is inserted to collect blood from the posterior vena cava.
- This procedure will be repeated three to four times to collect more volume of blood sample.

15. DISCUSSION

Blood collection from the experimental animals is one of the important procedures in biomedical research. Even a small error in the collection procedure may lead to a lot of variation in the results.

Points to be remembered

- Before starting any kind of blood sample collection, it must be ensured that all chemical, surgical, fluid requirements are available in the working site.
- Not more than two to three attempts should be made to collect any kind of *in vitro* biological sample (excluding biological secretion).
- The blood collection tube must be labeled before starting the experiment and blood sample collected in the appropriately labeled collection tube.

REFERENCES

1. Hoff J, Rlagt LV. Methods of blood collection in the mouse. Lab animals. 2000;29:47–53.
2. Blood sampling online. 2009. [12 screens]. Available from: http://www.nc3rs.org.uk/bloodsamplingmicrosite/page.asp?id=313[Last cited on 2010 Feb 24]
3. Guidelines for Survival Bleeding of Mice and Rats. NIH-ARAC Guidelines [online] 2005. Jan 12. Available from: http:// oacu.od.nih.gov/ARAC/Bleeding.pdf [accessed on 2010 Mar 09]
4. McGuill MW, Rowan AN. ILAR J. 1989. Biological Effects of Blood Loss: Implications for Sampling Volumes and Techniques; p. 31.
5. Procedure for rabbit blood collection [online] Available from:http://www.research.uky.edu/ori/univet/resources/sop/Procedure_rabbit_blood_collection.pdf [Last cited on 2010 Feb 23]
6. CPCSEA guidelines for laboratory animal facility. Indian J Pharmacol. 2003;35:257–74.
7. Vogel HG, editor. Drug discovery and evaluation: Pharmacological assays. 2nd ed. Berlin: Springer; 2002.

8. Anaesthesia and Analgesia in Laboratory Animals at UCSF [online] Available from:http://www.iacuc.ucsf.edu/Index.asp[Last cited on 2010 Feb 24]. and Available from:http://www.iacuc.ucsf.edu/Proc/awRatFrm.asp.[Last cited on 2010 Feb 24]

9. Hem A, Smith AJ, Solberg P. Saphenous vein puncture for blood sampling of the mouse, rat, hamster, gerbil, guinea pig, ferret and mink. Lab Anim. 1998;32:364–8.

10. Paulose CS, Dakshinamurti K. Chronic catheterization using vascular-access-port in rats: Blood sampling with minimal stress for plasma catecholamine determination. J Neurosci Methods. 1987; 22:141–6.

11. Yoburn BC, Morales R, Inturrisi CE. Chronic vascular catheterization in the rat: Comparison of three techniques. Physiol Behav. 1984;33:89–94.

CHAPTER 4

Diagnosis of Rabies via RT-PCR on Skin Samples of Wild and Domestic Animals

Ulrike Zieger

School of Veterinary Medicine, St. George's University,
St. George's, Grenada

ABSTRACT

In developing countries, brain tissues from rabies suspect animals are not always available for diagnosis for a variety of reasons, such as lack of transport to submit a carcass or the difficulty of removing an animal's head or brain under field conditions. To enable diagnosis in such cases, there is a need for a reliable method, using an alternative non-neural tissue, which can be removed and submitted to the diagnostic laboratory without special training or equipment. In human medicine, skin is used successfully for the detection of rabies virus antigen using RT-PCR technology. Little work has been done in animals using RT-PCR on skin or extracted hair follicles. The current study was conducted in Grenada on skin from 36 wild and domestic animals, in which rabies virus infection had been confirmed in brain tissue via the direct fluorescent antibody (DFA) test, and in 31 negative control animals. RT-PCR on skin yielded a sensitivity of 97.2% (35/36) and a specificity of 100% (31/31). It is concluded that the examination of skin samples via RT-PCR provides a valuable diagnostic alternative in those cases where brain tissue is not readily available.

KEYWORDS

Rabies Virus, Diagnosis, Skin, Animals, RT-PCR

1. INTRODUCTION

Rabies is a globally distributed infectious disease of the central nervous system of mammals, caused by members of the genus Lyssavirus in the family Rhabdoviridae. The rabies virus (RABV) is the type species of the genus Lyssavirus and causes most of the human rabies cases. Transmission typically occurs via the bite of an infected animal [1] . RABV antigen is detected in stained brain sections using the direct fluorescent antibody (DFA) test, which is the diagnostic gold standard [2] [3] .

In Grenada, a small island in the Caribbean Sea, rabies is endemic. It is maintained in a wildlife host, the small Indian mongoose (Herpestes auropunctatus), with spillover into other animals or humans [4] - [7] . Unfortunately, many rabies suspects are never submitted and carcasses are burned or buried without laboratory confirmation. Reasons given include the lack of adequate transport, especially where livestock is involved, or the inaccessibility by car of the area where an animal has died, quick decomposition of a carcass in the tropical climate, or lack of expertise to safely remove the head or brain. At other times, brain removal is delayed leading to late diagnostic reporting.

To overcome some of the above limitations, it would be of advantage if even laymen could collect an easily accessible non-neural tissue without having to open or transport an entire carcass for the reliable diagnosis of rabies virus infection. RABV moves centripetally along nerves from the site of entry to the brain, where it multiplies and begins to cause clinical signs. Then, it moves centrifugally along nerves to peripheral tissues [8] . Among these tissues, saliva, tear fluid, or corneal impressions have been widely used [9] , especially in human medicine, but viral shedding in these fluids is too inconsistent to give reliable results, unless taken as serial samples [10] . In contrast, examination of skin yielded very promising results: serial cryosections of skin from a variety of animals showed a sensitivity of 98% in the post-mortem rabies diagnosis using DFA when compared to the DFA test performed on brain tissue [11] . Limitations of using skin for DFA are that at least 20 sections need to be examined, the expensive equipment, and that decomposition renders tissues unsuitable for DFA testing. PCR can be used even on decomposed tissues, and a highly reliable PCR protocol was described using skin biopsies in human rabies cases [12] . Some data are available using PCR on animal skins [13] - [16] , yet reports are not promising: For example, Wacharapluesadee et al. [13] examined extracted whisker follicles and hair follicles of dogs that had died of rabies in Thailand and concluded that sensitivities of 81.8% using real-time PCR, and 66.7% using RT-PCR were "...not sufficient to help physicians (decide) whether to administer post exposure prophylaxis" in bite victims.

In the current study, we determined whether skin samples taken from animals after death and using conventional RT-PCR technology would be sufficiently sensitive in the diagnosis of RABV infection to serve as an easily obtainable alternative to brain tissue.

2. MATERIAL AND METHODS

2.1. Animals

A total of 36 confirmed rabid animals and 31 negative control animals were used. All cases had been submitted for rabies diagnosis to the Pathobiology Department, School of Veterinary Medicine, St. George's University, Grenada, West Indies, between March 2011 and May 2015. Carcasses were processed within 24 hours of an animal's death, or when this was not possible, carcasses were kept at −20˚C until processing.

2.2. Samples

To reduce the potential risk of saliva contamination of the skin, carcasses were thoroughly cleaned under running water. A 2 - 4 cm^2 section of skin was removed from the muzzle area, using a fresh sterile blade and taking care that the underlying muscle tissue or oral mucosa were not included. For a subset of animals, skin sections were also taken from the shoulder and hip area. All skin sections were individually stored dry at −20˚C in 50mL plastic containers until processing. Sampling and processing of saliva and brain samples from these animals including total RNA extraction, RABV antigen detection using DFA, RT-PCR, real-time PCR and sequencing have been described [7] .

2.3. RNA Extraction

For the extraction of total RNA, a skin sample was cut into sections of approximately 1 mm^3 using fresh sterile blades. Care was taken to cut the skin at oblique angles to increase exposure of hair follicles. Excess hair and fat were cut off and removed. Total RNA was extracted from approximately 30 mg of skin, using RNeasy Mini Kit spin column extraction (Qiagen GmbH, Hilden, Germany) following manufacturer's instructions. Skin was lysed in 0.6 mL RLT/beta-mercaptoethanol lysis buffer containing 1 mm zirconia/silica beads at a bead volume equal to skin volume (Biospec Products, Inc. Bartlesville, OK, USA), and milling samples in a beadbeater (Mini Beatbeater™, Biospec Products) for 20 sec, resting for 60 sec, and milling again for 20 sec. Samples were centrifuged after lysis at 13,000 rpm for 3 min and 0.5 mL of the clear lysate was used in the spin column protocol. The final elution volume was 30 µL.

Total extracted RNA was measured using Nanodrop 2000 (Thermo Scientific, Waltham, MA, USA). The minimum concentration used in this study was 2.3 ng RNA/µL and the maximum was 99.7 ng RNA/µL. Any sample containing more than 100 ng RNA/µL was diluted to a final concentration of 40 - 50 ng RNA/µL. Samples, which contained less than 1 ng RNA/µL or with a 260/280 ratio below 1.90 were discarded and the extraction repeated.

2.4. RT-PCR

RABV antigen was detected via RT-PCR as recently described [7] . Briefly, the one-step iScript protocol (Biorad Laboratories, Hercules, CA, USA) was used on 1 - 50 ng total RNA per 25 µL reaction mix. The primers used for amplification targeted a 110 bp fragment of the highly conserved region of the nucleoprotein (N)-gene: JW 12 (5'-ATGTAACACCYCTACAATG-3') and N 165-146 (5'-GCAGGGTAYTTRTACTCATA-3'). An Eppendorf Mastercycler ProS (Eppendorf AG, Hamburg, Germany) was used and RT-PCR products were visualized using 3 % agarose gel electrophoresis and ethidium bromide staining.

2.5. Statistical Methods

Results obtained from the new method, i.e. RABV antigen detection in skin by RT-PCR, were compared to the gold standard of rabies diagnosis, the viral antigen detection in brain tissue by DFA. Sensitivity, specificity, Positive Predictive Value (PVP), Negative Predictive Value (NPV) and confidence limits at 95% were calculated using MedCalc statistical software.

3. RESULTS

Skin samples were taken from the muzzle area of 36 rabid animals and 31 negative controls. The animal species and numbers are listed in Table 1. All cases had previously been confirmed as RABV positive, respectively negative, using DFA on brain tissue [7] . RT-PCR results from skin samples taken from the head are also presented in Table 1. In relation to the DFA test performed on brain tissue, the RT-PCR test performed on skin samples taken from the head of animals showed a sensitivity of 97.2% (95%Cl: 85.47% - 99.93%) and a specificity of 100% (95%Cl: 88.78% - 100.00%). The Positive Predictive Value (PVP) was 100% (95% Cl: 90.00% - 100.00%) and the Negative Predictive Value (NPV) was 96.88 (95% Cl: 83.78% - 99.92%).

Of the animals that tested RABV positive in skin taken from the head, only eight were also available for the collection of skin from their shoulder and hip areas: 3 dogs, 3 mongooses and 2 cats. In seven of these animals (87.5%), RABV was detected in all three skin areas. In one dog, only the skin from the head tested RABV positive, whereas skin taken from the shoulder and hip areas were RABV negative.

Testing skin for RABV antigen via spin column extraction and RT-PCR as described here took three hours from tissue lysis to gel visualization.

Table 1. Comparison of RABV detection via RT-PCR in skin taken from the head to DFA in brain tissue in rabies-infected and uninfected animals.

Animal species	Numbers	
	RT-PCR positive (skin)/DFA positive (brain)	RT-PCR negative (skin)/DFA negative (brain)
Mongoose (*Herpestes auropunctatus*)	20/20 [100%]	5/5 [100%]
Dog	10/11 [90.9%]	17/17 [100%]
Cat	4/4 [100%]	3/3 [100%]
Goat	1/1 [100%]	1/1 [100%]
Cow	-	1/1 [100%]
Bats (2 *Glossophaga longirostris*; 1 *Molossus molossus*)	-	3/3 [100%]
Opossum (*Didelphis marsupialis*)	-	1/1 [100%]
TOTAL	**35/36 [97.2%]**	**31/31 [100%]**

4. DISCUSSION

The current study tested the use of RT-PCR on skin samples as an alternative to DFA testing on brain tissue for the post mortem diagnosis of rabies. Skin samples collected from animals' heads showed a sensitivity of 97.2% and a specificity of 100% compared to the gold standard.

Skin as a potential non-neural tissue for rabies diagnosis has long been investigated. In human medicine, immunofluorescence (IF) techniques on skin biopsies typically taken from the nape of the neck have been successfully used in the post-mortem [17] and ante mortem rabies diagnosis [18] [19] . PCR technology has meanwhile replaced IF-staining of skin sections, and the collection of skin biopsies along with saliva and CSF samples and testing by PCR technology is now recommended as a routine measure in the ante-mortem diagnosis of rabies in humans [12] [20] - [23] . Indeed, Dacheux et al. [12] reported a sensitivity of 98% using hemi-nested PCR applied to human skin and proposed to use skin as an alternative to brain DFA testing for the confirmation of rabies before and after death.

In veterinary medicine, IF techniques on skin specimens of rabid animals showed promising results [8] [11] [24] - [26] , but were never routinely used, probably due to the expensive equipment needed for cryosectioning and the number of sections to be stained. Few reports are available which examine the use of skin using PCR techniques, possibly because the ante-mortem diagnosis is not as important in animals as in humans or because PCR with its inherent contamination issues has not been accepted as a primary method in rabies diagnosis. Wacharapluesadee et al. [13] concluded that their sensitivity of 81.8% using real-time PCR on extracted dog whisker follicles was insufficient. Studies conducted in India used skin from various animals before they died of rabies and detected RABV antigen in 9 of 13 (69%) by RT-PC and in 11 of 13 (84%) by real-time PCR [14] [15] . Kaw et al. [16] only observed a sensitivity of 57.1% for ante-mortem skin samples using RT-PCR and 71.4% using real-time PCR in animals.

It is interesting to note that the false negative case in the current study was a dog, which had died of paralytic rabies. Wacharapluesadee et al. [13] also found that most of their false-negative results originated from dogs that had died of paralytic rabies. Similarly, a high false-negative rate was reported for human paralytic cases when examining non-neural tissues [27] . This supports the idea that viral spread from the infected brain to peripheral tissues differs between paralytic and furious forms.

Considering the future possibility to use skin for ante-mortem diagnosis in animals, collection of biopsies from the head of a rabies suspect would only be safe under very tight physical restraint, with heavy sedation and/or brief anesthesia. Samples collected from the shoulder or hip area of an animal would be easier and safer to take. Although the sample size for comparing skin results was small (n = 8), the current study showed that RABV antigen could be detected in the skin of the head, shoulder and hip of a rabid animal. One of the 8 animals examined tested rabies positive using head skin, but had no detectable RABV antigen in its skin from the shoulder and hip regions. Contamination of the head skin with saliva in this case could be excluded, as its saliva tested RABV negative. The hair follicles of the head, lips, and especially whiskers are particularly well innervated [17] , and being closest to the brain, it is reasonable to assume that the virus is detected here earlier and perhaps in higher numbers than elsewhere.

5. CONCLUSION

In the post-mortem diagnosis of rabies in animals, testing skin tissue via RT-PCR provides a highly sensitive and rapid alternative to the examination of brain tissue via DFA in those situations where brain is not readily available, or where the removal or examination of brain is delayed. The method described here may also be useful for the ante-mortem diagnosis of rabies during an animal's quarantine or clinical work-up stage.

ACKNOWLEDGEMENTS

The assistance of Dr. Alfred Chikweto and Dr. Keshaw Tiwari, SGU-SVM Grenada, in taking skin samples is greatly appreciated. The author is grateful to Dr. Ravindra Sharma, SGU-SVM, Grenada, and to Dr. Dirk Werling, RVC London, UK, for reviewing the manuscript. The support of St. Gorge's University via research grants #10005 and #12015 is gratefully acknowledged.

COMPETING INTERESTS

The author declares that there is no competing interest.

REFERENCES

1. Charlton, K.M. (1988) The Pathogenesis of Rabies. In: Campbell, J.M. and Charlton, K.M., Eds., Rabies. Kluwer Academic Publishers, Boston, 101-150.

2. OIE-World Organization for Animal Health (2015) Manual of Diagnostic Tests and Vaccines for Terrestrial Animals: Chapter 2.01.13 Rabies (Version Adopted in May 2013). http://www.oie.int/manual-of-diagnostic-tests-and-vaccines-for-terrestrial-animals/

3. World Health Organization (WHO) (2015) WHO Expert Consultation on Rabies. WHO Press, Geneva. http://www.who.int/rabies/home_diagnosis/en

4. Everard, C.O.R., James, A.C. and DaBreo, S. (1979) Ten Years of Rabies Surveillance in Grenada. Bulletin of the Pan American Health Organization, 13, 342-353.

5. Everard, C.O.R. and Everard, J.D. (1992) Mongoose Rabies in the Caribbean. Annals of the New York Academy of Sci- ences, 653, 356-366.

6. Jonkers, A.H., Alexis, F. and Loregnard, R. (1969) Mongoose Rabies in Grenada. West Indian Medical Journal, 18, 167-170.

7. Zieger, U., Marston, D.A., Sharma, R., Chikweto, A., Tiwari, K., Sayyid, M., Louison, B., Goharrriz, K., Breed, A.C., Werling, D., Fooks, A.R. and Horton, D.L. (2014) The Phylogeography of Rabies in Grenada, West Indies, and Prospects for Control. PLoS Neglected Tropical Diseases, 8, Article ID: e3251.

8. Umoh, J.U. and Blenden, D.C. (1982) The Dissemination of Rabies Virus into Cranial Nerves and Other Tissues of Experimentally Infected Goats and Dogs, and Naturally Infected Skunks. International Journal of Zoonoses, 9, 1-11.

9. Rupprecht, C.E., Hanlon, C.A. and Hemachudha, T. (2002) Rabies Re-Examined. Lancet, 2, 327-343.

10. Trimarchi, C.V. and Nadin-Davis, S.A. (2007) Diagnostic Evaluation. In: Jackson, A.C. andWunner, W.H., Eds., 2nd Edition, Rabies, Academic Press, San Diego, 411-470.

11. Blenden, D.C., Bell, J.F., Tsao, A.T. and Umoh, J.U. (1983) Immunofluorescent Examination of the Skin of Rabies- Infected Animals as a Means of Early Detection of Rabies Virus Antigen. Journal of Clinical Microbiology, 18, 631- 636.

12. Dacheux, L., Reynes, J.M., Buchy, P., Sivuth, O., Diop, B.M., Rousset, D., Rathat, C., Jolly, N., Dufourcq, J.B., Nareth, C., Diop, S., Iehle, C., Rajerison, R., Sadorge, C. and Bourhy, H. (2008) A Reliable Diagnosis of Human Rabies Based on Analysis of Skin Biopsy Specimens. Clinical Infectious Diseases, 47, 1410-1417.

13. Wacharapluesadee, S., Tepsumethanon, V., Supavonwong, P., Kaewpom, T., Intarut, N. and Hemachudha, T. (2012) Detection of Rabies Viral RNA by TaqMan Real-Time RT-PCR Using Non-Neural Specimens from Dogs Infected with Rabies Virus. Journal of Virological Methods, 184, 109-112.

14. Bansal, K., Singh, C.K.K., Ramneek, Sandhu, B.S., Deka, D., Dandale, M. and Sood, N.K. (2012) Antemortem Diagnosis of Rabies from Skin: Comparison of Nested RT-PCR with TaqMan Real Time PCR. Brazilian Journal of Veterinary Pathology, 5, 116-119.

15. Bansal, K., Singh, C.K., Sandhu, B.S., Sood, N.K. and Dandale, M. (2014) Antemortem Diagnosis of Rabies from Skin by TaqMan Real Time PCR. Indian Journal of Animal Research, 48, 597-600.

16. Kaw, A., Singh, C.K., Ramneek, Deka, D., Sandhu, B.S., Awahan, S. and Sood, N.K. (2014). Comparison of Hair Follicles and Skin for Ante Mortem Detection of Rabies: A Molecular Approach. Indian Journal of Public Health Research and Development, 5, 165-169.

17. Bryceson, A.D.M., Greenwood, B.M., Warrell, D.A., Davidson, N.M., Pope, H.M., Lawrie, J.H, Barnes, H.J., Ballie, W.E. and Wilcox, G.E. (1975) Demonstration during Life of Rabies Antigen in Humans. Journal of Infectious Diseases, 131, 71-74.

18. Crepin, P., Audry, L., Rotivel, Y., Gacoin, A., Caroff, C. and Bourhy, H. (1998) Intravitam Diagnosis of Human Rabies by PCR Using Saliva and Cerebrospinal Fluid. Journal of Clinical Microbiology, 36, 1117-1121.

19. Warrel, M.J., Looareesuwan, S., Manatsathit, S., White, N.J., Phuapradit, P., Vejjajiva, A., Hoke, C.H., Burke, D.S. and Warrell, D.A. (1988) Rapid Diagnosis of Rabies and Post-Vaccinalencephalitides. Clinical and Experimental Immunology, 71, 229-234.

20. Smith, W.B., Blenden, D.C., Fuh, T.H. and Hiler, L. (1972) Diagnosis of Rabies by Immunofluorescent Staining of Frozen Sections of Skin. Journal of the American Veterinary Medical Association, 161, 1495-1501.

21. Brito, M.G., Chamone, T.L., da Silva, F.J., Wada, M.Y., de Miranda, A.B., Castilho, J.G., Carrieri, M.L., Kotait, I. and Lemos, F.L. (2011) Antemortem Diagnosis of Human Rabies in a Veterinarian Infected When Handling a Herbivore in Minas Gerais, Brazil. Revista do Instituto de Medicina Tropical de Sao Paulo, 53, 39-44.

22. Macedo, C.I., Carnieli Jr., P., Brandao, P.E., Travassos da Rosa, E.S., de Novaes-Oliveira, R., Galera-Castilho, J., Medeiros, R., Rocha-Machado, R., de Olivera, R.C., Carrieri, M.L. and Kotait, I. (2006) Diagnosis of Human Rabies Cases by Polymerase Chain Reaction of Neck-Skin Samples. The Brazilian Journal of Infectious Diseases, 10, 341-345.

23. Smith, J., McElhinney, L., Parsons, G., Brink, N., Doherty, T., Agranoff, D., Miranda, M.E. and Fooks, A.R. (2003) Case Report: Rapid Ante-Mortem Diagnosis of a Human Case of Rabies Imported into the UK from the Philippines. Journal of Medical Virology, 69, 150-155.

24. Blenden, D.C., Frost, J.W., Wachendoerfer, G. and Dorsey, C. (1980) Identification of Rabies Virus Antigen in the Skin of Foxes. Zentralblatt für Veterinarmedizin Reihe B, 27, 698-704.

25. Blenden, D.C. (1981) Rabies in a Litter of Skunks Predicted and Diagnosed by Skin Biopsy. Journal of the American Veterinary Medical Association, 179, 789-791

26. Ciuchini, F., Pestalozza, S., Buonavoglia, C. and Irsara, A. (1984) Evaluation of Immunofluorescence Staining of Skin for the Diagnosis of Rabies. Zentralblatt für Veterinarmedizin Reihe B, 31, 32-35.

27. Wacharapluesadee, S. and Hemachudha, T. (2010) Ante- and Post-Mortem Diagnosis of Rabies Using Nucleic Acid Amplification Tests. Expert Review of Molecular Diagnostics, 10, 207-218.

CHAPTER 5

Genetic Diversity of PRRS Virus Collected from Air Samples in Four Different Regions of Concentrated Swine Production during a High Incidence Season

Barbara Brito [1,*], Scott Dee [2], Spencer Wayne [2], Julio Alvarez [1] and Andres Perez [1]

[1]Department of Veterinary Population Medicine, College of Veterinary Medicine, University of Minnesota, 385A Animal Science Veterinary Medicine Building, 1988 Fitch Avenue, St. Paul, MN 55108, USA

[2]Pipestone Veterinary Services, Pipestone, MN 56164, USA

ABSTRACT

Porcine Reproductive and Respiratory Syndrome (PRRS) is one of the most relevant swine diseases in the US, costing the industry millions of dollars per year. Unfortunately, disease control is difficult because of the virus dynamics, as PRRS virus (PRRSV) can be transmitted by air between farms, especially, in regions with high density of swine operations. While long distance airborne transport of PRRSV has been reported, there is little information regarding the dynamics of PRRSV airborne challenge in concentrated regions. The objective of this study was to describe the frequency of detection, dose and diversity of PRRSV in air samples collected across four concentrated production regions during the PRRS-high risk season in the Midwestern US (October–

December) in 2012. Between 29% and 42% of the air samples were positive in all four sampling sites. Sequencing of the recovered virus showed a wide diversity of field and vaccine variants. Higher frequency, dose, and diversity of PRRSV were observed in air at locations with higher pig density. These findings suggest that regional spread of PRRSV due to aerosol transmission of PRRSV represents a significant risk to susceptible herds in concentrated regions of domestic pig production where PRRSV is endemic.

KEYWORDS

porcine; Porcine Reproductive and Respiratory Syndrome; virus; livestock diseases; air filtration; Molecular Epidemiology

1. INTRODUCTION

Porcine Reproductive and Respiratory Syndrome (PRRS) virus (PRRSV) was first identified in 1991, and is now present in many countries worldwide [1,2] PRRSV is one of the primary causes of decreased swine production with losses estimated at 664 million dollars per year to the US industry [3]. PRRSV is a member of the genus *Arterivirus* [4], a single stranded positive sense enveloped RNA virus with a genome containing nine open reading frames (ORF) that code for seven structural proteins and 14 non-structural proteins [5].

Control and elimination of PRRSV in the US is challenging due to the high concentration of swine production in specific areas of the country along with an absence of official regulation, such as reporting or mandatory control actions [6]. Another aspect complicating PRRS control is the ability of the virus to be transported over extended distances by contaminated aerosols [7,8,9]. While data exist regarding the ability of the virus be transported via aerosols up to 9.2 m [9], little is known about the dynamics of daily PRRSV airborne challenge directly at the farm level. If the level of airborne PRRSV challenge on a day-to-day basis could be documented, decisions regarding the application of intervention strategies capable of reducing the airborne risk (e.g., area vaccination or air filtration) would be easier to make [10,11].

Therefore, in order to document the risk of PRRSV airborne challenge, the objective of this study was to measure frequency of detection and the dose and the diversity of PRRSV observed in daily air samples collected around four regions that differed in farm and pig density. Novel techniques were employed, including liquid cyclonic collectors for recovering virus from air [12] and Bayesian phylogenetic analysis of sequenced virus [13]. Characterization of virus detected in air samples will help guiding decisions regarding biosecurity measures in swine farms.

2. MATERIALS AND METHODS

2.1. Air Sampling Locations

Air samples were collected in the surroundings of four commercial sow units. The study farms were located in Eastern South Dakota (farm A), Northeast Iowa (farm B), Southwest Minnesota (farm C), and Northwest Iowa (farm D). Approximately 90% of the neighboring farms were finishers or wean-to finish type. County pig density was obtained from the National Agricultural Statistics Service census of Agriculture statistics 2012 census [14]. The four sow farms have similar, strict biosecurity protocols, including air filtration and introduction of gilts from PRRS-negative multipliers. Three of the four farms had no PRRS positive animals during the study period. In location D, three animal samples were positive to vaccine strain virus, and one to RFLP 1-26-2 virus.

To visualize the farm density in each of the four locations, neighboring pig farms within a 10 km radius, were mapped using a kernel density, implemented in an open source GIS software [15]. The radius set for the kernel density (influence of a farm in a circular area) was set to 5 km, using a quadratic bi-weight kernel shape.

2.2. Air Sample Collection and Quality Control Methods

For the purpose of air collection, a sampling kit was sent to each of the four study farms. The kit consisted of a plastic tote containing a liquid cyclonic collector (Midwest MicroTek, Brooking, SD, USA), sterile 7.2% saline (Nova-Tech, Grand Island, NE, USA), a notebook for record keeping, permanent markers, individually wrapped sterile Dacron swabs (Fisher Scientific, Waltham, MA, USA), 3 mL plastic Falcon tubes (Becton Dickinson, Franklin Lakes, NJ, USA), a box of disposable gloves, a spray bottle of disinfectant (0.8% Synergize, Preserve International, Atlanta, GA, USA) and paper towels.

For the purpose of quality control, prior to shipping to the farm, each item in the kit, along with its plastic container was swabbed and tested for the presence of PRRSV RNA by PCR. From October 15 through December 15, 2012, a period of time described as the peak epidemic season for PRRS in the region [16], daily air samples were collected outside of the facility from 7–7:30 AM each day. A liquid cyclonic collector was placed 30 m from the facility in the direction of the daily prevailing wind and samples were collected as described elsewhere [12]. To maximize consistency and quality of the sampling process, a designated individual on each farm conducted the sampling and handled all air samples. Those four individuals had been trained in using the sampling equipment, the proper handling of samples and record keeping using validated protocols. Following each daily collection, samples were stored on farm in an external freezer at −20 °C and the collection equipment was sanitized and swabbed to document the absence of residual PRRSV RNA. Finally, to validate the specificity of the sampling process, air samples were collected on an abandoned farm site. This site housed no swine and the closest swine farm to this site was 15 km away. For the purpose of air collection, a wall fan from an abandoned building was turned on for 30 minutes and a sample collected. Ten such samples were collected in October, November and December 2012 for a total of 30 samples. Following completion of the project period, all air samples from all farms were sent to the University of Minnesota Veterinary Diagnostic Laboratory (UMN-VDL) for testing.

Should positive samples be detected, it was planned to re-test all PCR positive samples and to re-sequence a subset (25%) of the samples to insure accuracy of laboratory methods. Finally, to insure that accidental sample contamination, secondary to contact with serum or oral fluids samples did not occur at the farm, at the Pipestone Veterinary Clinic

(PVC), or at the UMN-VDL, all samples were tested for PRRSV antibodies by ELISA [17].

2.3. Polymerase Chain Reaction

All samples collected during the study were tested for the presence of PRRSV RNA using a one-step TaqMan PCR assay (Perkin-Elmer Applied Biosystems, Foster City, CA, USA) at the UMN-VDL. The PCR assay detected the ORF 6 region of the North American (NA) PRRSV and the ORF 7 region of the European (EU) PRRSV [18]. Primers and probes employed were as follows:

PRRS NA/EU primers:

a) NA F (ORF 6) Forward primer (26 oligomer) (5'⇒3': sequence: GTAGTYGCRCTCCTTTGGGGGTGTA)

b) NA F2 (ORF 6) Forward primer (33 oligomer) (5'⇒3': sequence: TTCATCACYTCCAGRTGCCGTTTGTGCYTGCTA)

c) NA R (ORF 6) Reverse primer (18 oligomer) (5'⇒3': sequence: CGASAAATGCGTGGTTAT)

d) EU (ORF 7) Forward primer (19 oligomer) (5'⇒3': sequence: TGGCCAGCCAGT AATCAA)

e) EU (ORF 7) Reverse primer (21 oligomer) (5'⇒3': sequence: TGTGGCTTCTCAGGCTTTTTC)

PRRS NA/EU FAM primers labeled ORF6 TAMRA probe (22 oligomer)

a) (5'⇒3': FAM-TACATTCTGGCCCCTGCCCAYC-TAMRA) PRRS EU FAM labeled ORF7 TAMRA probe (25 oligomer)

b) (5'⇒3': 6FAM-TGCAATGATAAAGTCCCAGCGCCAG-TAMRA).

2.4. Virus Titration

All PCR-positive air samples were titrated using Marc-145 cells and MEM supplemented with 8% fetal calf serum, antibiotics and antifungal agents, to quantify the level of infectious virus in the sample [19,20].

2.5. Sequencing

To estimate genetic diversity of PRRSVs recovered from PCR-positive air samples to those collected from field cases, the ORF 5 region of the PRRSV was sequenced using previously published techniques [21].

2.6. Phylogenetic Analysis

Sequences collected during this study, in addition to 105 sequences available from PVC records of PRRS-infected animals collected from October 15 to December 15, 2012 (study period) were also included. Furthermore 216 sequences from the same database including different viruses collected from 2002 to 2013 were added to increase the robustness of the phylogeny reconstruction.

All sequences were aligned using MUSCLE [22]. The nucleotide substitution and codon partition schemes were selected using the PartitionFinder software package [23]. The best partition scheme and nucleotide substitution model were selected based on the Bayesian information criteria (BIC). We reconstructed the phylogeny of the sequences by simulation and sampling of trees using a Markov chain Monte Carlo MCMC algorithm implemented in BEAST v1.8 [24]. We compared relaxed uncorrelated exponential and uncorrelated lognormal clock models. For the tree prior we compared the coalescent constant population size and the coalescent Bayesian skyline.

We simulated the trees for 500,000,000 generations and checked that the effective sample size value reached >100 in all parameters. The convergence and mixing of estimates of parameters generated in the simulations were visualized and evaluated in Tracer v1.6 [25]. To determine if a strict clock would be an appropriate assumption for the data instead of the relaxed clock, the distribution of the estimated coefficient of variation of the relaxed models was examined, if 90% of the observations did not include 0, the strict model was ruled out. All four models were run (each relaxed clock with either coalescent constant or Bayesian skyline population). The final model was selected using the posterior simulation-based analogue of Akaike's information criterion (AICM), which we run 100 times (AICM) [26].

To visualize the tree file generated, the tree was annotated burning all simulations before reaching convergence according to the observed trace. The tree was visualized using Figtree v.1.5 [27].

2.7. Diversity of Sequences Found in the Sampling Areas

We compared two approaches to characterize the diversity of PRRSV sampled from air. One of the measures was the traditionally used pairwise percentage of genetic identity and the phylogenetic distance estimated by the maximum clade credibility (MCC) tree estimated by Bayesian methods previously described. For this purpose we computed the sequence identity between all pairs of sequences (number of identical positions divided by the total length of the alignment) using BioEdit [28]. The phylogenetic distance estimated by Bayesian methods was measured as the cophenetic distance between all sequences pairs of the MCC, which was obtained using ape package in R. [29]. The distance between the sequences in this phylogenetic tree, reflect the sum of time of evolution to a common ancestor of each pair of sequences, being a measure that takes into account the estimated relationship of the virus based on its divergence in time. On the other hand, the pairwise percentage of identity only indicates the number of nucleotides in which two aligned sequences differ. The correlation between the percentage identity and cophenetic distance values was computed using Pearson's coefficient (R), to evaluate the association between those two genetic distance methods.

Histograms of pairwise genetic differences obtained by the two methods were constructed for sequences collected at each air sampling location.

We further classified different viruses as defined by the phylogenetic tree clades. A clade was defined as a group of viruses reaching a common ancestor in 2009 or earlier in time. Summary statistics of sequences identities for each of the clades were also computed.

For each of the sample locations, the frequency of RNA recovery (positive samples divided by total samples collected), and the amount of virus (in $TCID_{50}$/mL) were recorded. Locations were compared in terms of proportion of positive samples and virus concentration using z-test for two independent proportions and ANOVA, respectively.

3. RESULTS

The description of each of the areas in terms of farm density within 10 km, and pig density within the county are shown in Table 1. The isopleth map generated using a kernel smoothing process (Figure 1) shows the density of farms within 10 km of each air sampling location. Farm C was within the lowest farm density area, including 20 neighboring farms (Table 1). Farms A and B both had 42 neighbors within 10 km, but in farm A, neighbors were clustered in half of the 10 km circle area, in contrast with farm D, which had its neighbors homogenously distributed throughout all the 10 km radius. Farm D was evidently the one within the higher farm-density-area, having 99 neighbors total throughout all surroundings within the 10 km radius. Although location A was within a higher farm-density area than location C, the county-level pig density of location C was 17 times higher than location A (Table 1).

3.1. Air Samples

Across all four sampling areas 37% (80/217) of the air samples collected were PCR positive, with frequency of positive samples in each area ranging from 29% to 42% (Table 1). When analyzed by z-test comparison of independent proportions, the differences of the number of PRRSV PCR-positive air samples across the four sampling areas were not statistically significant ($p > 0.05$). The mean quantity of viable PRRSV in air samples ranged from $10^{3.8}$ to $10^{5.3}$ $TCID_{50}$/mL. When analyzed by ANOVA, the mean viral titer detected in air samples around the farm D was statistically different than those observed across the other three sites ($p < 0.01$). Pertaining to quality control, all samples from collection kits prior to farm shipment and upon returning from farms were PCR-negative. All samples collected from cyclonic collectors following sanitation at the end of the sampling day across all four sites were PCR negative. In regards to air sample quality control, all repeat testing of samples by PCR (n = 80) and sequencing (n = 20) indicated identical results as the initial testing. No antibodies against PRRSV were detected in any samples, suggesting that contamination of samples during collection and/or processing did not take place. Finally, all samples (n = 30) collected from the farm site free of swine were negative for PRRSV by PCR.

3.2. Phylogenetic Analysis

A total of 68 air samples were sequenced and used in the analysis, of which 14, 12, 16 and 25 were from sampling locations A, B, C and D, respectively.

For model selection, the strict clock was ruled out. Based on the lowest AICM the relaxed uncorrelated exponential clock with a constant population size tree prior was selected. The MCC phylogenetic tree constructed with the selected model is shown in Figure 2. The sequences collected for the study were clustered in 19 clades according to the criteria of having a common ancestor within the five years prior to the study period. The mean percentage of paired sequence identity within each of these clusters ranged between 94.5% and 98.9%.

Clade 3 and 5 correspond to sequences closely related to modified live virus (MLV) variants, associated with RFLP 2-5-2 (MLV vaccine) and 1-4-2 (ATP vaccine). Vaccine-related sequences were detected in air samples from all four air-sampling locations.

In each sampling location, different viruses were collected. Sequences from sampling locations A, B, C and D were distributed in different clusters, some of them closely related to sequences obtained from pig samples in different clades (Figure 2). In locations A, B, and C we found similar number of viral strains as determined by a different clade (nine, nine, and eight, respectively), whereas there was a higher viral diversity in air samples from location D (14 different strains).

Table 1. Summary of virus collection frequency, concentration, and strain diversity in each of the air sampling collection sites.

Farm	Number neighboring farms (10 km)	County density (pigs/km^2)	Mean (SD) distance of the farms within 10 km	Frequency (+) samples (%)	Mean viable virus in log_{10} $TCID_{50}$/mL (range)	Number of virus sequenced	Number of virus strains *
A	42	13	7.53 (1.33)	15/51 (0.29)	4.1 (1.0–5.6)	14	9
B	42	163	6.39 (2.27)	22/53 (0.42)	4.4 (2.0–5.0)	12	9
C	20	221	6.19 (2.46)	18/54 (0.33)	3.8 (1.0–5.3)	16	8
D	99	477	6.95 (2.17)	25/59 (0.42)	5.3 (3.6–6.5)	25	14

* Strains defined by phylogenetic tree in Figure 2.

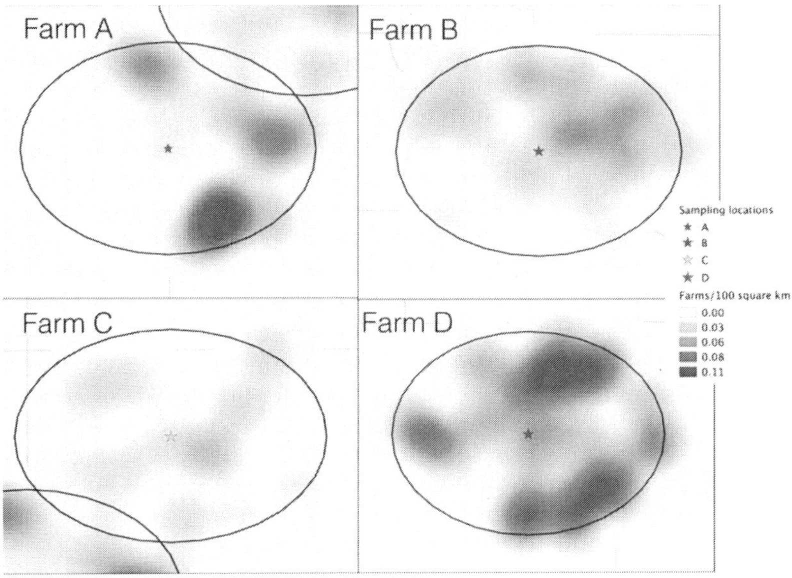

Figure 1. Farm density kernel within 10 km of each of the air sample locations: Farm A (South Dakota), farm B (Northeast Iowa), farm C (Southwest Minnesota), farm D (Northwest Iowa).

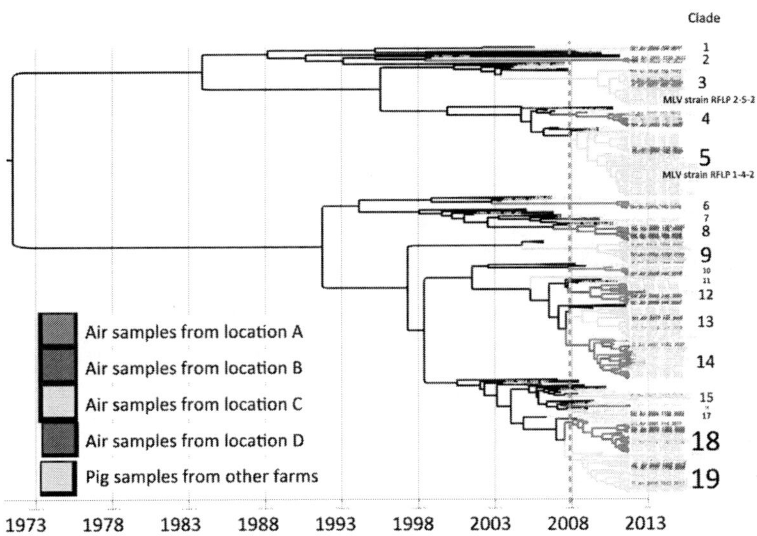

Figure 2. Maximum clade credibility tree. Sequences collected from air samples locations are indicated by color, location A (red), B (blue), C (green) and D (purple). The dotted line represents the cutoff to determine phylogenetic clades.

3.3. Comparison of Pairwise Sequence Identity and Genetic Distance in the Phylogenetic Tree

A scatter plot of the pairwise identity of the 173 sequences included in the phylogeny (air samples and animal samples), and the distance calculated from the phylogenetic tree is shown inFigure 3. There was a moderate correlation (R = 0.7) between the percentage of identity and cophenetic distance between each pair of sequences. That finding suggests that there was a discrepancy in some pairs of sequences in which the distance computed by percentage of identity and the phylogenetic relationship (cophenetic distance computed from the estimated phylogeny),*i.e.*, sequences with high percentage of identity, but for which the evolutionary distance was relatively high.

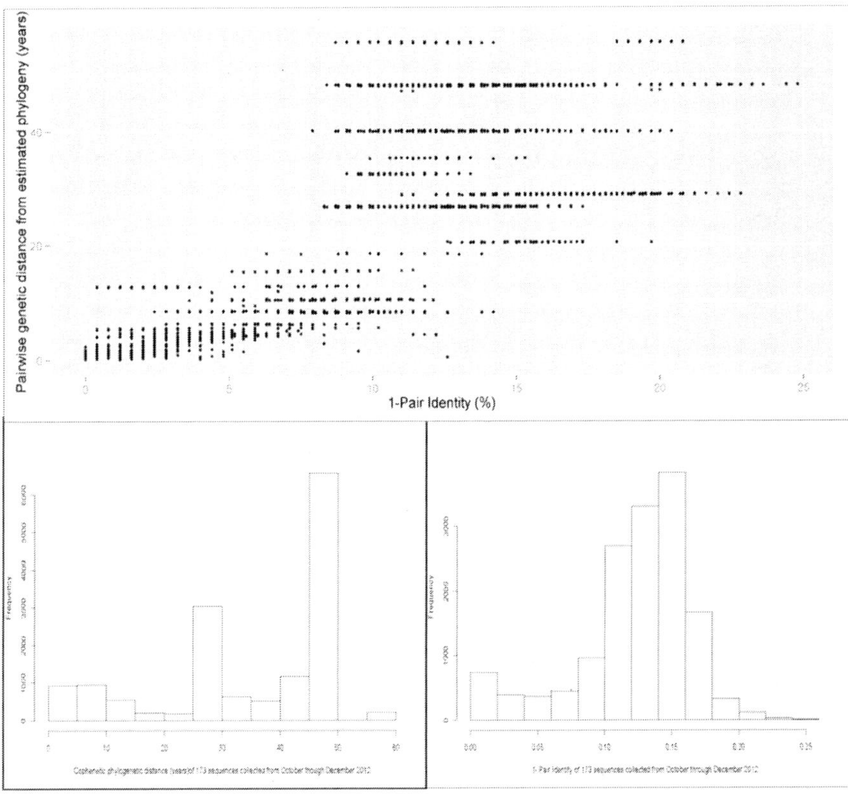

Figure 3. (**A**) Scatter plot of pairwise genetic identity and the distance computed from the phylogenetic tree. The total number of pairs of the 173 sequences was 14,878. Pearson's correlation was 0.70 (95%CI 0.71-0.69). (**B**) and (**C**) Histograms of distribution of pairwise identity and tree cophenetic distance values between pairs of sequences.

The distribution of the genetic distance measured by the estimated phylogenetic tree shows that approximately half of the pair distances were between 0 and 40 years (sum of time of evolution to the most recent common ancestor), and half of the sequence pairs had a genetic distance between 45 and 50 years. The cophenetic distance between a pair of sequences in this phylogenetic tree can be interpreted approximately as twice the value estimated to reach the common ancestor (because sampling of sequences was done within a narrow window of time relative to the total evolutionary time), for example, if a pair of sequences have a cophenetic distance value of 50, then the time to the most recent common ancestor of that pair of sequences would be ~25 years. In contrast, the histogram of the distribution of pair identities showed values centered between 85% and 90% identity (Figure 3). These values in the center of the paired identity distribution account for most of the genetic distance classified differently by the cophenetic distance method, as seen in the correlation matrix.

4. DISCUSSION

This study was conducted in an effort to estimate the dynamics of airborne challenge in areas with a high density of swine production. For this purpose, we evaluated the frequency of detection, concentration and diversity of PRRSV in air samples collected around four commercial swine farms. Our results showed that in all four locations, diverse populations of PRRSV variants were present in the air, from eight to 14 different strains in one location. In addition, as hypothesized, as the density of swine production increased in a region, the frequency of detection, and concentration of virus in the air was higher. Therefore, the results of this study document that the PRRSV airborne challenge to farms in regions of concentrated swine production is significant and emphasizes the need for developing a strategy to reduce this risk when herd biosecurity plans are being designed.

In the Midwestern US, there are numerous areas with high farm and pig density, as the one described for farm D. The study suggests a very high risk of the virus entering the farm via aerosol, specifically in this study, between October and December, when there is a higher incidence of the disease. Although less aerosolized virus was detected in location C, which had the least farm-density (but high county pig density), the concentration and frequency of virus detected is also relevant to consider as a source of infection for surrounding farms.

While some previous studies have also found that airborne transmission of PRRSV as an important route of viral spread [7,9] others have suggested that the relevant route of spread is animal movement, based on molecular epidemiology [30,31,32]. In theory, if the only route of PRRS spread were airborne, an association between the space and genetic variation would be expected, but because the virus is spread by animal movement, as well as air, evidence of this association is difficult to assess. Animal movement data and complete sequence information of PRRSV in farms within an area would help understand these dynamics and adjust for virus incoming by introduction of new animals.

Another interesting finding of the study is the numerous PRRSV strains found in the air. Even in the low-farm-density areas, at least eight different viruses were sequenced. This highlights that many PRRSV strains have the ability to spread through air. This is also important, because vaccination or acclimation of gilts account for some strains, but may not protect against already circulating, or strains recently introduced in the area. This consideration, again should be more strongly considered for those areas with higher farm and pig-density, where more virus strains may circulate exemplified by the 14 different viruses present in location D.

An interesting issue arises by the fact that MLV virus was consistently isolated from air samples. If this virus reaches by air a negative farm, or another farm, it may lead to a positive PCR or ELISA result. This finding highlight the need to sequence the virus found in a farm, and may not rule out vaccine virus in a farm that does not vaccinate.

The negative results of all controls in our study support the robustness of the results obtained. Another strength of this study is the assessment of viral diversity using two different methods: computation of the pairwise identity as well as the genetic distance determined by the phylogenetic tree constructed by Bayesian methods.

Bayesian phylogenetic analysis has been used to study and understand the evolution of RNA viruses, such as Avian Influenza, HIV, among many others [33,34,35]. These methods have been popular over distance-based methods (*i.e.*, Neighbor-Joining), to estimate phylogenies of viruses for different reasons. Distance based methods only use absolute distances between pairs of aligned sequences, which can result in an underestimation of the true genetic distance if some nucleotide sites have gone through multiple substitutions events. This feature may explain, at least in part, the moderate correlation ($R = 0.7$) of results obtained with both methods in the study here. Bayesian methods also allow analyzing serially sampled data (the date of collection of the sample is taken into account), so the estimation of the branches lengths represents substitution per site per time period rather than just substitution per site values. It also allows computing complex models, not subject to assumptions that may not be biologically plausible (e.g., accounting for rate of heterogeneity among sites or relaxed clock models). Finally, in contrast to distance-based methods, Bayesian analysis search for the tree that best fits the data across different potential trees, and also estimates the uncertainty of the phylogeny.

The results of our study suggest that aerosol transmission of PRRSV between farms may likely be a highly relevant route of infection in swine dense regions endemic for PRRSV. We detected several viral strains circulating in the air within swine production areas, with a distinctive high concentration in the location with the highest farm and pig-density. One interesting observation is that strains from clades 1, 2 and 17 (Figure 2) were not similar to other viruses isolated from pigs included in this study. Apparently the frequency of these strains in animals is low (or alternatively may not present in local farms), but detected by environmental air sampling.

Implementation of biosecurity measures that prevent airborne transmission of PRRSV in high density areas would not only protect from PRRSV, but there is also evidence that endemic influenza can also be transmitted through aerosol, as well as exotic devastating diseases that could affect swine production, such as foot-and-mouth disease and African swine fever [36,37].

5. CONCLUSION

In conclusion, we found concentrated virus with high frequency, as well as a high diversity of strains of PRRSV in air samples collected around commercial sow farms in swine-dense regions. Concentration, frequency and number of strains were higher in areas where farm and pig densities were higher. The study suggests that aerosolized virus may likely be an important route of transmission between farms in high-density areas. Results also suggest that computing pairwise identity between sequences pairs, as a measure of genetic distance, may not always reflect the true evolutionary distance. These results will contribute to the understanding of the dynamics of PRRSV spread and, ultimately, the prevention and control of one of the most devastating diseases of swine.

ACKNOWLEDGMENTS

This study was funded in part by grants from Pipestone Applied Research, USDA NIFA Award #2008-55620-19132, the Minnesota MN Drive program, and the University of Minnesota CVM Population Systems Signature Program and Ag Experiment Station Research Funds.

AUTHOR CONTRIBUTIONS

Conceived and designed the study: SD, AP, BB, SW, JA. Air sample collection: SD, SW. Wrote the paper: BB, SD. Analyzed the data: BB, SD.

CONFLICTS OF INTEREST

The authors declare no conflict of interest.

REFERENCES AND NOTES

1. Wensvoort, G.; Terpstra, C.; Pol, J.M.; ter Laak, E.A.; Bloemraad, M.; de Kluyver, E.P.; Kragten, C.; van Buiten, L.; den Besten, A.; Wagenaar, F.; *et al*. Mystery swine disease in the Netherlands: The isolation of Lelystad virus. *Vet. Q.* **1991**, *13*, 121–130.

2. Albina, E. Epidemiology of porcine reproductive and respiratory syndrome (PRRS): An overview. *Vet. Microbiol.* **1997**, *55*, 309–316.

3. Holtkamp, D.J.; Kliebenstein, J.B.; Neumann, E.J.; Zimmerman, J.J.; Rotto, H.F.; Yoder, T.K.; Yeske, P.E.; Mowrer, C.L. Assessment of the economic impact of porcine reproductive and respiratory syndrome virus on United States pork producers. *J. Swine Health Prod.* **2013**, *21*, 72–84.

4. Cavanagh, D. Nidovirales: A new order comprising Coronaviridae and Arteriviridae. *Arch. Virol.* **1997**, *142*, 629–633.

5. Dokland, T. The structural biology of PRRSV. *Virus Res.* **2010**, *154*, 86–97.

6. Corzo, C.A.; Mondaca, E.; Wayne, S.; Torremorell, M.; Dee, S.; Davies, P.; Morrison, R.B. Control and elimination of porcine reproductive and respiratory syndrome virus. *Virus Res.***2010**, *154*, 185–192.

7. Dee, S.; Otake, S.; Oliveira, S.; Deen, J. Evidence of long distance airborne transport of porcine reproductive and respiratory syndrome virus and Mycoplasma hyopneumoniae.*Vet. Res.* **2009**, *40*, 39.

8. Lambert, M.E.; Arsenault, J.; Poljak, Z.; D'Allaire, S. Correlation among genetic, Euclidean, temporal, and herd ownership distances of porcine reproductive and respiratory syndrome virus variants in Quebec, Canada. *BMC Vet. Res.* **2012**, *8*, 76.

9. Otake , S.; Dee, S.; Corzo, C.; Oliveira, S.; Deen, J. Long-distance airborne transport of infectious PRRSV and Mycoplasma hyopneumoniae from a swine population infected with multiple viral variants. *Vet. Microbiol.* **2010**, *145*, 198–208.

10. Dee, S.; Cano, J.P.; Spronk, G.; Reicks, D.; Ruen, P.; Pitkin, A.; Polson, D. Evaluation of the long-term effect of air filtration on the occurrence of new PRRSV infections in large breeding herds in swine-dense regions. *Viruses* **2012**, *4*, 654–662.

11. Linhares, D.C.; Cano, J.P.; Wetzell, T.; Nerem, J.; Torremorell, M.; Dee, S.A. Effect of modified-live porcine reproductive and respiratory syndrome virus (PRRSv) vaccine on the shedding of wild-type virus from an infected population of growing pigs. *Vaccine* **2012**, *30*, 407–413.

12. Pitkin, A.N.; Deen, J.; Dee, S.A. Use of a production region model forevaluation of routes of transmission and protocols of biosecurity forPRRS virus. *Vet. Microbiol.* **2009**, *136*, 1–7.

13. Yang, Z.; Rannala, B. Bayesian phylogenetic inference using DNA sequences: A Markov chain Monte Carlo method. *Mol. Biol. Evol.* **1997**, *14*, 717–724.

14. National Agricultural Statistic Service, United States Department of Agriculture. Census of Agriculture. Available online: http://quickstats.nass.usda.gov/ (accessed on 11 November 2014).

15. QGIS Development Team. Geographic Information System. Open Source Geospatial Foundation Project. 2014. Available online: http://qgis.osgeo.org (accessed on 11 November 2014).

16. Tousignant, S.; Morrison, B. PRRS incidence/prevalence pilot study, 2012. In Proceedings of the 2012 Allen D. Leman Swine Conference, St. Paul, MN, USA, 15–18 September 2012; Volume 39. Available online: http://conservancy.umn.edu/ handle/ 11299/ 139370 (accessed on 11 November 2014).

17. Kittawornrat, A.; Prickett, J.; Wang, C.; Olsen, C.; Irwin, C.; Panyasing, Y.; Ballagi, A.; Rice, A.; Main, R.; Johnson, J.; *et al.* Detection of Porcine reproductive and respiratory syndrome virus (PRRSV) antibodies in oral fluid specimens using a commercial PRRSV serum antibody enzyme-linked immunosorbent assay. *J. Vet. Diagn. Invest.* **2012**, *24*, 262–269.

18. Egli, C.; Thür, B.; Liu, L.; Hofmann, M.A. Quantitative TaqMan RT-PCR for the detection and differentiation of European and North American strains of porcine reproductive and respiratory syndrome virus. *J. Virol. Methods* **2001**, *98*, 63–75.

19. Kim, H.S.; Kwang, J.; Yoon, I.J.; Joo, H.S.; Frey, M.L. Enhanced replication of porcine reproductive and respiratory syndrome (PRRS) virus in a homogenous subpopulation of MA-104 cells. *Arch. Virol.* **1993**, *430*, 477–483.

20. Reed, L.J.; Muench, H. A simple method of estimating fifty percent endpoints. *Am. J. Hyg.***1938**, *27*, 493–497.

21. Murtaugh, M.P.; Elam, M.R.; Kakach, L.T. Comparison of the structural protein coding sequences of the VR-2332 and Lelystad virus strains of the PRRS virus. *Arch. Virol.* **1995**, *140*, 1451–1460.

22. Edgar, R.C. MUSCLE: multiple sequence alignment with high accuracy and high throughput. *Nucleic Acids Res.* **2004**, *32*, 1792–1797.

23. Lanfear, R.; Calcott, B.; Ho, S.Y.; Guindon, S. Partitionfinder: Combined selection of partitioning schemes and substitution models for phylogenetic analyses. *Mol. Biol. Evol.* **2012**, *29*, 1695–1701.

24. Drummond, A.J.; Rambaut, A. BEAST: Bayesian evolutionary analysis by sampling trees. *BMC Evol. Biol.* **2007**, *7*, 214.

25. Rambaut, A.; Drummond, A.J. Tracer v1.6: MCMC trace analyses tool. Available online: http://beast.bio.ed.ac.uk/Tracer (accessed on 11 November 2014).

26. Raftery, A.E.; Newton, M.A.; Satagopan, J.M.; Krivitsky, P.N. Estimating the integrated likelihood via posterior simulation using the harmonic mean identity. In *Bayesian Statistics*; Bernardo, J.M., Bayarri, M.J., Berger, J.O., Eds.; Oxford University Press: Oxford, UK, 2007; pp. 1–45.

27. Rambaut, A. FigTree v1.4: Tree figure drawing tool. Available online: http://tree.bio.ed.ac.uk/software/figtree/ (accessed on 11 November 2014).

28. Hall, T.A. BioEdit: A user-friendly biological sequence alignment editor and analysis program for Windows 95/98/ NT. *Nucleic Acids Symp. Ser.* **1999**, *41*, 95–98.

29. Paradis, E.; Claude, J.; Strimmer, K. APE: Analyses of phylogenetics and evolution in R language. *Bioinformatics* **2004**, *20*, 289–290.

30. Badaoui, B.; Grande, R.; Calza, S.; Cecere, M.; Luini, M.; Stella, A.; Botti, S. Impact of genetic variation and geographic distribution of porcine reproductive and respiratory syndrome virus on infectivity and pig growth. *BMC Vet. Res.* **2013**, *9*, 58.

31. Kwong, G.P.S.; Poljak, Z.; Deardon, R.; Dewey, C.E. Bayesian analysis of risk factors for infection with a genotype of porcine reproductive and respiratory syndrome virus in Ontario swine herds using monitoring data. *Prev. Vet. Med.* **2013**, *110*, 405–417.

32. Rosendal, T.; Dewey, C.; Friendship, R.; Wootton, S.; Young, B.; Poljak, Z. Spatial and temporal patterns of porcine reproductive and respiratory syndrome virus (PRRSV) genotypes in Ontario, Canada, 2004–2007. *BMC Vet. Res.* **2014**, *10*, 83.

33. Lemey, P.; Rambaut, A.; Drummond, A.; Suchard, M. Bayesian phylogeography finds its root. *PLoS Comput. Biol.* **2009**, *5*, e1000520.

34. Volz, E.M.; Koelle, K.; Bedford, T. Viral Phylodynamics. *PLoS Comput. Biol.* **2013**, *9*, e1002947.

35. Grenfell, B.T.; Pybus, O.G.; Gog, J.R.; Wood, J.L.; Daly, J.M.; Mumford, J.A.; Holmes, E.C. Unifying the epidemiological and evolutionary dynamics of pathogens. *Science* **2004**, *16*, 327–332.

36. De Carvalho Ferreira, H.C.; Weesendorp, E.; Quak, S.; Stegeman, J.A.; Loeffen, W.L.A. Quantification of airborne African swine fever virus after experimental infection. *Vet. Microbiol.* **2013**, *165*, 243–251.

37. Gloster, J.; Sellers, R.F.; Donaldson, A.I. Long distance transport of foot-and-mouth disease virus over the sea. *Vet. Rec.* **1982**, *110*, 47–52.

CHAPTER 6

Potential bacterial core species associated with digital dermatitis in cattle herds identified by molecular profiling of interdigital skin samples

Martin W. Nielsen, Mikael L. Strube, Anastasia Isbrand, Worood D.H.M. Al-Medrasi, Mette Boye, Tim K. Jensen, Kirstine Klitgaard*

National Veterinary Institute, Technical University of Denmark, Bülowsvej 27, 1870 Frederiksberg C, Denmark

ABSTRACT

Although treponemes are consistently identified in tissue from bovine digital dermatitis (DD) lesions, the definitive etiology of this debilitating polymicrobial disease is still unresolved. To study the microbiomes of 27 DD-infected and 10 healthy interdigital skin samples, we used a combination of different molecular methods. Deep sequencing of the 16S rRNA gene variable regions V1–V2 showed that *Treponema, Mycoplasma, Fusobacterium* and *Porphyromonas* were the genera best differentiating the DD samples from the controls. Additional deep sequencing analysis of the most abundant genus,*Treponema,* targeting another variable region of the 16S rRNA gene, V3–V4, identified 15 different phylotypes, among which *Treponema phagedenis*-like and *Treponema refringens*-like species were the most abundant. Although the presence of *Treponema*spp., *Fusobacterium necrophorum* and *Porphyromonas levii* was confirmed by fluorescence *in situ* hybridization (FISH), the results for *Mycoplasma* spp. were inconclusive. Extensive treponemal epidermal infiltration, constituting more than 90% of the total bacterial population, was observed in 24 of the 27 DD samples. *F. necrophorum*and *P. levii* were superficially located in the epidermal lesions and were present in only a subset of samples. RT-qPCR analysis showed that treponemes were also actively expressing a panel of virulence factors at the site of infection. Our results further support the hypothesis that species belonging to the genus *Treponema* are major pathogens of DD and also provide sufficient clues to motivate additional research into the role of *M. fermentans, F. necrophorum* and *P. levii* in the etiology of DD.

KEYWORDS

Bovine digital dermatitis; Next generation sequencing; 16S rRNA; Microbiome;*Treponema*; RT-qPCR

1. INTRODUCTION

Bovine digital dermatitis (DD) is presently the most common cause of lesions associated with lameness in dairy cattle worldwide (Laven and Logue, 2006). This debilitating disease has a huge impact on animal welfare and productivity (Losinger, 2006). DD most probably has a bacterial etiology because no virus or fungus of any significance has been identified from the lesions (Krull et al., 2014). Likewise, the immunological response of afflicted cows and the positive effect of antibiotic treatment seem to support this notion (Demirkan et al., 1999). Various species of spirochetes have been cultivated from the affected lesions (Pringle et al., 2008 and Evans et al., 2009); and with the advent of 16S rRNA gene sequencing and fluorescence *in situ* hybridization (FISH), it has been shown that a considerable number of *Treponema* phylotypes are present in DD lesions (Moter et al., 1998, Klitgaard et al., 2008 and Rasmussen et al., 2012). Phylogenetically, the identified treponemes mainly belong to 4 major groups, *Treponema phagedenis*-like,*Treponema denticola*/*Treponema pedis*-like, *Treponema medium*/ *Treponema vincentii*-like and *Treponema refringens*-like phylotypes (Yano et al., 2010, Santos et al., 2012 and Klitgaard et al., 2013). These bacteria are consistently reported to be located deep within the affected tissue (in the front part of the lesion) and to outnumber other bacterial morphotypes. Furthermore, their presence is always associated with degenerative and necrotic changes of the infected tissue (Choi et al., 1997). These tissue-destructive properties have been reported to be enhanced by and to probably be dependent upon the presence of other microorganisms found in these polymicrobial infections (Edwards et al., 2003). However, it must be emphasized that Koch's postulates have not been fulfilled for DD, and the definitive etiology of the disease is therefore still unresolved.

In addition to treponemes, the range of taxa identified from DD lesions by bacteriological, histopathological, and molecular biological investigations include the genera*Porphyromonas*, *Prevotella*, *Fusobacterium*, *Campylobacter*, *Bacteroides*,*Mycobacterium*, *Mycoplasma* and *Guggenheimella* (Wyss et al., 2005, Santos et al., 2012 and Krull et al., 2014). Many of these bacterial groups, however, seem to be primarily located in the superficial parts of the lesions, and their potential roles in the pathogenesis of DD are still unresolved. The bacterium *Dichelobacter nodosus*, the cause of bovine foot rot, is commonly seen invading the epidermis in DD lesions (Rasmussen et al., 2012).

The aim of the present study was to determine whether other bacteria, besides treponemes, are consistently associated with DD and to investigate whether we could link distinct phylotypes of *Treponema* to specific stages of DD lesion development as previously indicated (Krull et al., 2014 and Zinicola et al., 2015). FISH analysis was applied to visualize and localize selected microbial taxa identified by deep sequencing of the 16S rRNA gene. Finally, the expression of a number of known or putative *Treponema*virulence genes in DD lesions was measured by reverse transcriptase real-time PCR (RT-qPCR). To date, only the immunological host gene response has been investigated for DD (Zuerner et al., 2007), and the active expression of candidate virulence genes would be a further indication that *Treponema* are actively involved in the etiology of this disease.

2. MATERIALS AND METHODS

2.1. Sample collection and preparation

All sample material was collected from a Danish cattle slaughterhouse. Tissue samples were collected from the hind feet of slaughtered Holstein-Friesian dairy cattle after routine slaughter procedures. The feet were gently rinsed with water and cleaned with a soft-haired brush to remove any gross organic debris, and the dorsal/posterior skin between the digits was classified as early (M1), acute ulcerative (M2), healing (M3) or chronic lesions (M4) according to Döpfer et al. (1997).

Tissue samples were collected with a sterile scalpel and transferred to a sterile Petri dish. The samples were subdivided and stored for histologic examination and DNA purification. The sample portions intended for 16S rRNA gene analysis were immediately transferred to a nucleic acid stabilization solution (RNA*later*®, Ambion, Austin TX) and stored at −20 °C. Samples intended for histopathology analysis were fixed in 10% neutral buffered formalin solution, dehydrated, embedded in paraffin wax, sectioned (5-µm thick) and mounted on SuperFrost Plus slides (Menzel-Gläser, Braunschweig, Germany).

2.2. Histopathology

All tissue samples were stained with hematoxylin and eosin (H&E) for histopathological evaluation. The degree of epidermal damage and the inflammatory response of the dermis were scored from 0 to 2 according to Rasmussen et al. (2012), with minor modifications. A score of 0 was defined as normal epidermis or mild epidermal damage with some epithelial proliferation and hyperkeratosis, dermis being unaffected. Score 1 was defined as epithelial proliferation and acanthosis (with ballooning degeneration and mal-keratinization) while the inflammatory response was seen as mild to moderate increase in the number of lymphocytes and mononuclear cells. Score 2 was defined as severe epithelial proliferation, acanthosis and exudation, erosion or necrosis of the dermal papilla and moderate to severe infiltration with lymphocytes and/or mononuclear cells in dermis and perivascular dermatitis.

2.3. Fluorescent in situ hybridization (FISH)

The oligonucleotide probes used in this study are listed in Table S1 and include probes specific for Domain Bacterium, *Treponema* spp., *Fusobacterium necrophorum,Dichelobacter nodosus*, *Porphyromonas levii*, *Mycoplasma* spp., and *Mycoplasma fermentas*. Newly designed oligonucleotide probes were selected using the software ARB (http://www.arb-home.de). The oligonucleotide probes were 5′ labeled with fluorescein isothiocyanate (FITC) or the isothiocyanate derivative Cy3 (Eurofins MWG Operon, Ebersberg, Germany). The hybridization was carried out at 45 °C with 100 µl of hybridization buffer (10 µl of 1 M Tris [pH 7.2], 18 µl of 5 M NaCl, 1 µl of 10% sodium dodecyl sulfate, 71 µl of H_2O) and 500 ng of each probe for 16 h in a Sequenza slide rack (Thermo Shandon, Cheshire, United Kingdom). The sections were then washed three times in prewarmed (45 °C) hybridization buffer for 9 min and subsequently washed three times in prewarmed (45 °C) washing solution (10 µl of 1 M Tris [pH 7.2], 18 µl of 5 M NaCl, 72 µl of H_2O). The sections were rinsed in water, air dried, and mounted in Vectashield (Vector Laboratories Inc., Burlingame, CA) for epifluorescence microscopy. An Axioimager M1 epifluorescence microscope equipped for epifluorescence with a 100-W HBO lamp and filter sets 43 and 38 was used to visualize Cy3 and FITC, respectively. Images were obtained using an AxioCam MRm version 3 FireWiremonocrome camera and AxioVision software, version 4.5 (Carl Zeiss, Oberkochen, Germany).

2.4. Nucleotide extraction

DNA and RNA were extracted from the biopsies using an AllPrep® DNA/RNA/miRNA kit (Qiagen, Hilden, Germany). Portions (20 mg) of stabilized tissue were first homogenized in 600 µl of RTL plus buffer (included in the kit). A sterile 5-mm steel bead (Qiagen) was added, and samples were run two times on a TissueLyser II (Qiagen) at 20 Hz for 2–5 min per run. All the subsequent procedures were performed according to the protocols of the supplier. After RNA extraction, the material was additionally treated with TURBO™ DNase, according to the protocol provided by the manufacturer (Ambion), to ensure total DNA degradation in the RNA preparations. For the evaluation of DNA and RNA concentration and purity, samples were analyzed on a NanoDrop ND-1000 spectrophotometer (Fisher Scientific, Wilmington, MA). DNA samples with $A260/A280$ ratios >1.5 were selected for further analysis. The RNA quality was estimated on an Agilent Bioanalyzer (Agilent Technologies Inc., Santa Clara, CA).

2.5. Reverse transcription

Reverse transcription was performed using a QuantiTect Reverse Transcription kit according to the manufacturers instructions (Qiagen).

The concentration and quality of cDNA samples were measured with a NanoDrop ND-1000 spectrophotometer (Fisher Scientific). All cDNA samples with *A*260/*A*280 ratios ≥1.8 were included in the downstream analysis.

2.6. Preparation of 16S rRNA gene amplicon libraries and sequencing

PCR amplification of DNA was accomplished with a universal bacterial primer set (Wilmotte et al., 1993) and a *Treponema*-specific primer set (Klitgaard et al., 2008) targeting the V1–V2 region and the V3–V4 hyper variable regions of the 16S rRNA gene (primer sequences listed in S1). Each sample was amplified with unique forward and reverse primers that included an added hexamer barcode at their 5′ ends. Amplification PCRs were performed in 50-µl reaction mixtures containing 5 µl of 10xPCR Gold Buffer (Applied Biosystems, Foster City, CA, USA) 1.5 mM $MgCl_2$ solution (Applied Biosystems), 200 µM of each deoxynucleoside triphosphate (Amersham Biosciences, Piscataway, NJ), 0.4 µM of each specific primer, 2.5 U of AmpliTaq Gold DNA polymerase (Applied Biosystems), and 2 µl of template DNA. For both primer sets, thermal cycling using a T3 thermocycler (Biometram, Göttingen, Germany) was performed as follows: denaturation at 94 °C for 6 min; 30 cycles of denaturation at 94 °C for 45 s, annealing at 57 °C for 45 s, and extension at 72 °C for 90 s. A final elongation step of 10 min was followed by cooling to 4 °C. Positive (DNA) and negative (dH_2O) controls were included for each PCR setup. The DNA concentration and quality of the PCR amplicons from all samples were assessed with an Agilent 2100 Bioanalyzer (Agilent Technologies Inc.) prior to high-throughput sequencing (data not shown). Equal amounts of all amplicons were pooled (final concentration between 3.8–4 µg) and purified with the Qiagen Mini Elute kit (Qiagen) according to the protocol of the manufacturer. The DNA was submitted to the National High-Throughput DNA Sequencing Centre at the University of Copenhagen, Denmark for sequencing on the Illumina MiSeq™ platform.

Sequences generated by Illumina MiSeq are available under the accession number SRP065522 in the NCBI Sequence Read Archive (SRA).

2.7. Sequence analysis

For both sets of sequences, the obtained reads were analyzed using the BION-meta software (http://box.com/bion). The de-multiplexing step was performed according to the primer- and barcode sequences. Forward and reverse sequences were joined, allowing no gaps, a maximum mismatch percentage of 80% and an overlap length of minimum 35 bp. Next, the sequences were cleaned at both ends by the removal of bases of a quality less than 98%, which is equivalent to a Phred score of 17. Identical sequences were de-replicated into consensus sequences of 300–322 bp. Consensus sequences of at least 250 nucleotides in length were mapped into a table, according to the individual barcodes, and taxonomically classified against the Ribosomal Database Project database II (RDP II; http://rdp.cme.msu.edu/index.jsp) using a word length of 8 and a match minimum of 90%. To allow for the comparison of relative abundance between samples, the number of reads for each barcode was normalized.

To elucidate the microbial patterns involved in DD-etiology, multivariate analysis of the resulting microbial profiles was carried out by Analysis Of Similarities (ANOSIM) and Constrained Analysis of Principal coordinates (CAP) with Bray-Curtis distances constrained by histological score (Anderson and Willis, 2003). Genera identified by CAP as being associated with DD were further analyzed with the Kruskal-Wallis test in order to pinpoint differences in bacterial compositions between scores 1 and 2.

2.8. Real-time quantitative reverse transcriptase PCR (RT-qPCR)

Amplification, detection and real-time quantitative analysis were performed using the Rotor-Gene Q (Qiagen, Hilden, Germany). The sequences of the *Treponema* genes studied were obtained from GenBank, and the primers were designed in Primer3 (http://biotools. umassmed. edu/bioapps/primer3_www.cgi) (Rozen and Skaletsky, 2000). The nucleotide sequences of the primers used in this study are listed in Table S1. Each PCR was performed in a 25-µl reaction mixture containing 12.5 µl of QuantiTect SYBR Green PCR master mix (Qiagen), a primer concentration of 0.2 µM (a mix of primers targeting *T. denticola*, *T. vincentii* and *T. phagedenis*) and 200 ng of cDNA. Three replicas were included for each sample. The thermal cycling conditions were as follows: 15 min at 95 °C, followed by 40 cycles of 30 s at 94 °C, 20 s at 55 °C, and 20 s at 72 °C. Data collection was performed during each extension phase. Positive controls (DNA) and negative controls (distilled water) were included in each run. The control for DNA contamination was performed by RT-qPCR on the mRNA samples. Melting curve analysis was performed for each species of *Treponema*, which resulted in single product-specific melting curves for all primer sets.

For every gene, a standard curve comprising a pool of cDNA (5 points of 5 × dilutions) from all the samples was included in each run and was used to determine the amplification efficiency and the target concentration of the individual samples. The reference genes were tested in geNorm (GenEx 6; MultiD Analyses AB, Göteborg, Sweden), and the geometric mean of two internal reference genes, *tpiA* and *tkt,* was used to correct the raw values for the genes of interest (Vandesompele et al., 2002).

2.9. PCR of *fermentans*

All the included DNA samples and a subset of cDNA samples were subjected to selective detection of *M. fermentans* by the PCR amplification of a conserved 206 bp region of the insertion sequence-like element, IS*1550*, of *M. fermentans* (Wang et al., 1992) and a 414 bp region of the macrophage-activating lipopeptide gene (*malp*) (Afshar et al., 2008). The 206 bp region was detected by standard PCR with 2.5 U thermostable recombinant AmpliTaq DNA polymerase (Applied Biosystems), 5 µl of 10xPCR Buffer II (Applied Biosystems), 1.5 mM $MgCl_2$ Solution (Applied Biosystems), 100 µM of each deoxynucleoside triphosphate (Amersham Biosciences, Piscataway, NJ), 0.2 µM of each specific primer, and 0.3 µg template DNA in a 50 µl total volume. Thermal cycling was performed in a T3 thermocycler (Biometra, Göttingen, Germany) under the following conditions: denaturation at 94 °C for 3 min and 45 cycles of denaturation at 94 °C for 45 s, annealing at 55 °C for 45 s, and extension at 72 °C for 90 s. A final elongation step of 10 min was followed by cooling to 4 °C. For detection of the *malp* gene, we first amplified the entire MALP-404 lipoprotein coding sequence. From this 1373 bp PCR fragment, a second nested PCR was performed using primers targeting a 414 bp fragment. PCR and cycling conditions were performed according to the protocol provided by Afshar et al. (2008). For all PCR reactions, boiling lysate of *M. fermentans* (E-9) was used as a positive control. The PCR products were separated on a 2% E-gel (Invitrogen, Carlsbad, CA, USA), and the DNA was visualized by UV fluorescence. Two malp PCR products were sequenced on an ABI 3130 genetic analyzer (Applied Biosystems) using a BigDye Terminator v3.1 cycle sequencing kit (Applied Biosystems), according to the manufacturer's instructions.

3. RESULTS

3.1. Pathology and histology

Twenty-seven cases of DD with varying degrees of affected skin and 10 control cases were selected and sampled at a slaughterhouse. Four cases were grossly characterized by circumscribed ulcerations and hyperkeratosis with crust formation (M3). Fourteen cases showed circumscribed lesions dominated by significant hyperkeratosis and proliferation of the epidermis but without visible ulceration (M4). Nine cases showed only mild focal hyperkeratosis/proliferation of the epidermis (not definable according to Döpfer et al. (1997); thus, in this study, they were designated M5). Ten cases revealed no gross lesions. The scores are listed in Table 1.

The samples in the present project originated from a slaughterhouse; we had no clinical anamnesis or information regarding the chronology of the lesions nor was it possible to pinpoint the infection stage from the histological examinations with certainty. Instead, the lesions were classified into groups based on histological criteria. The histology scores are presented in Table 1. Seventeen samples were histopathologically characterized by severe epithelial proliferation and acanthosis with exudation, erosion or necrosis of the dermal papilla (score 2) as illustrated in Fig. 1. Ten samples were characterized by epithelial proliferation and acanthosis with varying ballooning degeneration and mal-keratinization (score 1). All cases without gross lesions appeared within the normal range (score 0).

Table 1. Summary of pathology and histology scores for the 37 bovine skin samples.

	Pathology score		Fluorescence in situ hybridization (FISH)[b]			
Sample no.	Döpfer score[a]	Histology	*Treponema*	*P. levii*	*F. necro-phorum*	*D. nodosus*
1	m4	1	3	0	3	0
2	m3	1	3	0	1	0
3	m4	1	3	0	0	0
4	m4	2	3	0	0	0
5	m4	2	3	0	2	0
6	m4	2	3	2	2	0
7	m4	2	2	1	3	3
8	m4	2	3	0	2	3
9	m4	1	3	0	0	0
10	m4	1	0	0	0	1
11	m4	2	3	0	0	0
12	m3	2	3	3	0	0
13	m4	2	3	3	0	0
14	m4	2	3	0	0	0
15	m4	2	3	2	0	0
16	m3	2	3	2	0	2
17	m4	2	3	2	0	1
18	m5	1	ne	ne	ne	ne
19	m5	1	3	0	0	1
20	m5	1	3	0	1	2
21	m5	1	0	0	0	0
22	m3	2	3	0	0	2
23	m5	2	3	1	1	1
24	m5	2	3	0	0	1
25	m5	1	3	0	0	1
26	m5	2	2	0	0	3
27	m5	2	3	0	0	0
C1-C10		0	0	ne	ne	ne

[a] for explanation of this score see in text.
[b] Due to the inconclusive results, the FISH data for *M. fermentans* were not included in the Table.

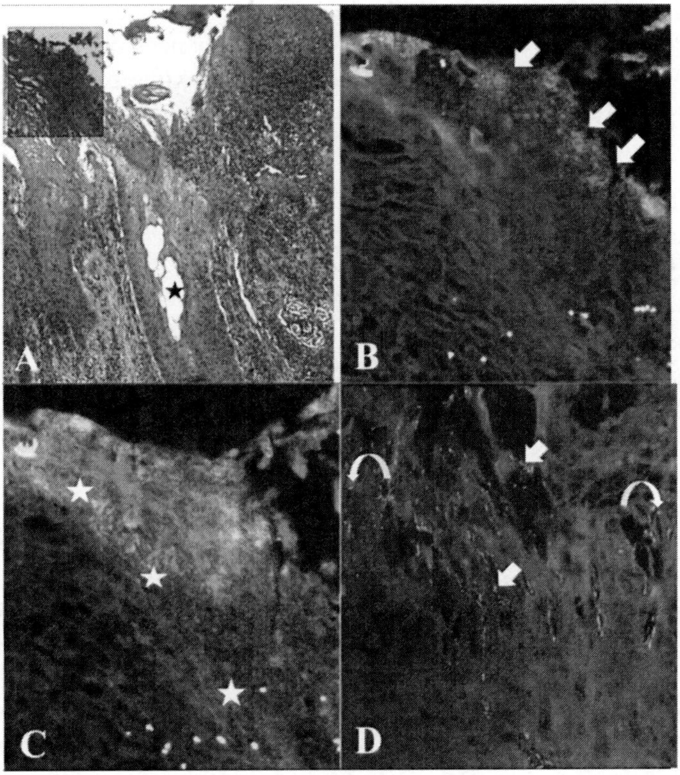

Figure 1. (A) Bovine digital dermatitis lesion (biopsy No. 13), characterized by acanthotic epidermis with degenerated ballooning keratinocytes, exocytosis, ulceration of dermal papillae, and bacterial infiltration. Hair follicle (star). The shading represents the area used for fluorescent in situ hybridization (FISH) on parallel sections in B and C, H&E. B) Demonstration of *Porphyromonas levii* organisms (orange)(arrows) in the superficial layers of the lesion, FISH. C) Demonstration of severe infiltration with*Treponema spp.*

3.2. 16S rRNA sequencing

We sequenced 37 samples with the primers V1–V2 and 31 samples with the primers V3–V4. After de-multiplexing according to the sequences of the barcodes and primers, 5,047,554 and 3,235,052 sequences remained in the V1–V2 pool and V3–V4 pool, respectively. The 3′ and 5′ ends of these sequences were further trimmed according to quality. Sequences with quality below 98% were discarded. In total, 2,428,737 (V1–V2 pool) and 1,613,962 (V3–V4 pool) joined sequences were used for taxonomic classification, equivalent to average reads per sample of 65,641 and 52,063, respectively. Of these sequences, 27% of the V1–V2 pool and 76% of the V3–V4 pool were taxonomically classifiable according to the RDPII database (http://rdp.cme.msu.edu/).

3.3. Bacterial amplicons

The dataset included sequences of microbiota isolated from 37 bovine hoof skin samples, including 10 from macroscopically normal skin (controls) and 27 from visible lesions with variable morphology and lesion scores (Table 1). The consensus sequences were taxonomically assigned using the RDP II database. Diversity of the three histological groups was estimated by using a Shannon index (Hill et al., 2003). We observed a significant decrease in phylotype richness in the DD lesions (histology score 1 ($H' = 1.02 \pm 0.43$) and 2 ($H' = 1.17 \pm 0.73$)) when compared to normal skin (histology score 0 ($H' = 3.53, \pm 0.23$)). The phyla in the control samples mainly belonged to *Firmicutes*, *Actinobacteria*, *Proteobacteria* and *Bacteroidetes* (Fig. 2A), together covering 15 families with an average relative abundance of more than 1%. Of these, only four families,*Corynebacteriaceae*, *Carnobacteriaceae*, *Ruminococcaceae*, and *Moraxellaceae*, exceeded an average relative abundance of 5% (Fig. 2B).

The sequences originating from the 27 DD lesions could largely be ascribed to the phyla*Spirochaetes*, *Tenericutes*, *Fusobacteria* and *Bacteroidetes* (Fig. 2A). At the family level, only *Spirochaetaceae* (47%), *Mycoplasmataceae* (28%) *Fusobacteriaceae* (8%) and *Porphyromonadaceae* (7%) exceeded an average relative abundance of 5% (Fig. 2B). Within these families the main genera represented were *Treponema*, *Mycoplasma*, *Fusobacterium* and *Porphyromonas*. Fig. 3 shows the individual distribution of the identified amplicons at the genus or species level in the 27 DD-biopsies. In 15 of the samples, *Treponema* species accounted for over half of the total sequence population detected. The second most frequent bacterial species, *M. fermentans*, had an abundance of 20% in 16 samples.

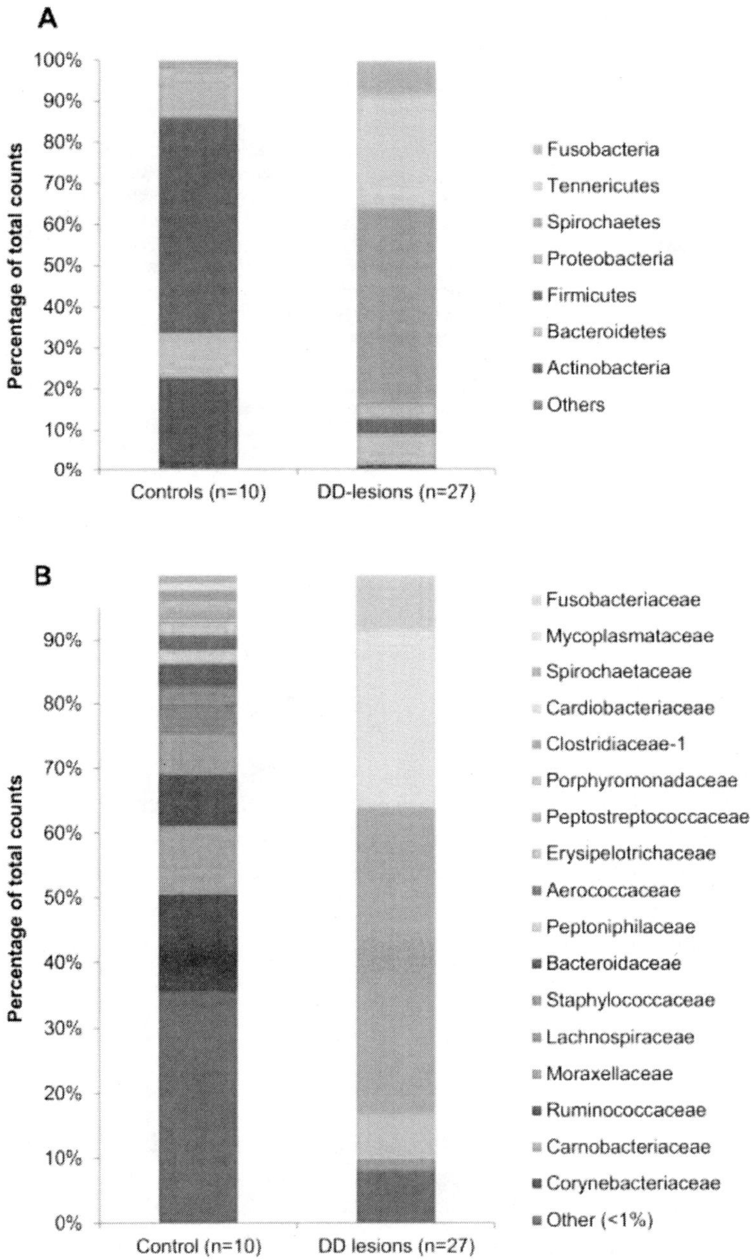

Figure 2. Average relative abundance of taxonomic bacterial groups in control and DD lesions derived from 37 biopsies using 16S rRNA gene based metagenomics and V1–V2 general eubacterial primers. A) At the phylum level, B) at the family level (relative abundance of bacterial families that represented at least 1% of the bacterial reads acquired).

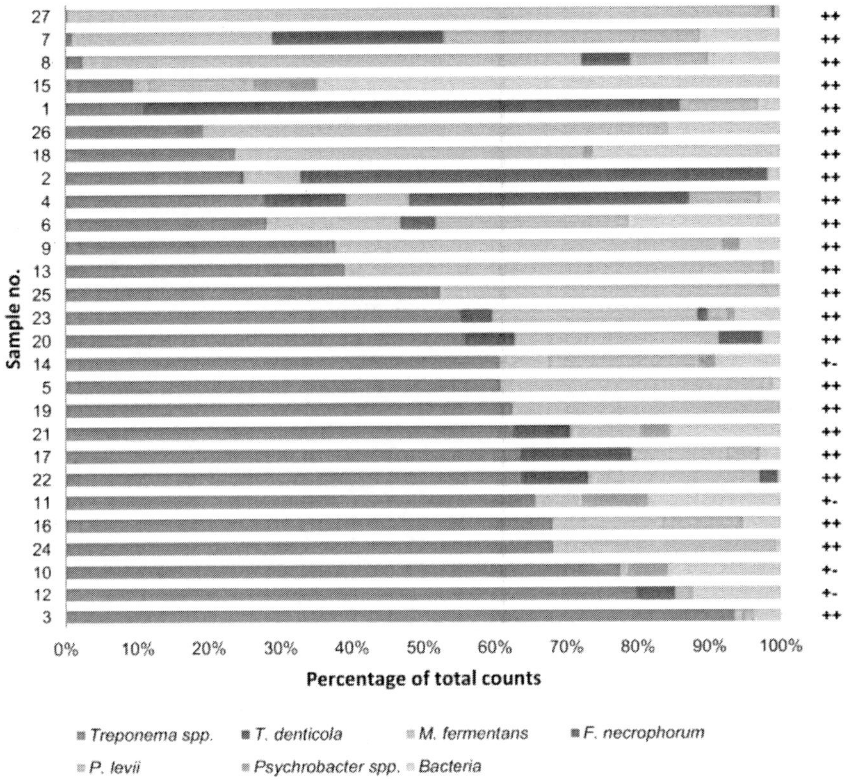

Figure 3. Relative abundance of bacterial genera/species in the microbiota of bovine hoof skin samples with macroscopic lesions, sorted by the prevalence of *Treponema*. Sequence amplicons were derived from 27 biopsy specimens by the use of V1–V2 bacterial primers and 16S rRNA gene-based metagenomics. To the right of the bars, the results of the *M. fermentans*-specific PCRs; First row +/−: result of IS*1550* PCR, second row +/−: result of *malp* PCR.

Analysis of similarities (ANOSIM) revealed a significant difference in the microbial composition between samples from the control group (histology score 0) and the DD-biopsies (histology score 1 and 2), but not between the DD-biopsies with different histology scores. We used constrained analysis of principal coordinates (CAP) to probe the underlying microbial patterns associated with DD etiology on the genus level. In this analysis, 49.5% of the variation could be explained. In the first dimension, accounting for 47.8% of the variation, *Treponema, Mycoplasma, Fusobacterium* and *Porphyromonas*were the genera that best differentiated the DD samples from the controls. On the second dimension, explaining very little of the variation, i.e., 1.7%, we observed a limited separation of histological scores 1 and 2, with lesion 2 scores associated with higher abundance of *Porphyromonas* and *Mycoplasma* and lower abundance of *Treponema*and *Fusobacterium*. In the univariate analysis, however, only the distribution of*Porphyromonas* was significantly different between scores 1 and 2 ($p < 0.01$, Kruskal-Wallis, Fig. S1).

3.4. *Fermentans*-specific PCRs

The presence of *M. fermentans* in the biopsies was confirmed by two species-specific PCR assays targeting the IS*1550* and *malp* genes of this bacterium. Whereas all controls were negative with the *malp* PCR, C2 and C7 were positive with the IS*1550* PCR. The IS*1550* PCR is extremely sensitive and should be able to detect one single copy of the gene (Wang et al., 1992). This result may explain why C2 and C7 tested positive; we did see *M. fermentans* amplicons in these samples, although in very low amounts (below <0.04% abundance). The PCR results for the DD biopsies are shown in Fig. 3. To determine whether IS*1550* and *malp* were also expressed by *M. fermentans* in the DD lesions, the two PCR assays were likewise performed with cDNA as template for a subset of samples (no. 1, 5, 8, 9, 12, 18, 19, 17, 24, 25 and 27). No *M. fermentans* amplicons were detected from these samples, which indicated no or very limited expression of IS*1550* and *malp* in the infected tissue.

3.5. *Treponema* amplicons

In contrast to the sequences generated with the V1–V2 general primers, with which only*T. denticola* could be determined to the species level, we were able to identify approximately 94% of the amplicons generated with the *Treponema*-specific primers to the species/phylotype level. *Treponema* phylotypes with a relative abundance of <1% were joined in a group named *Treponema* spp. In total, 15 different species/phylotypes of*Treponema* were identified from 31 samples. The number of phylotypes with a relative abundance of >1% in the individual samples ranged between 1 and 10, with an average of 5 per sample. Six of the control samples, C4-C9, were omitted from this part of the study as no PCR amplicons could be generated from these samples with the *Treponema*-specific primers.

Fig. 4 illustrates the relative distribution of *Treponema* phylotypes in the individual samples. Very few DD-related treponemes were seen in control samples C1 and C2. The treponemes from the DD biopsies and control samples C3 and C10 were all homologous to species previously identified in DD lesions. The most common species was *T. phagedenis* (24%), followed by the *T. refringens*-like and not-yet-cultivable *PN-20* (15%), PT1 (14%) and PT4 (10%). There appeared to be no obvious correlation between lesion score and the relative abundance of *Treponema* variants. Because *Treponema* specific sequences could be amplified from only a subset of the controls, no multivariate analysis was performed on the V3–V4 generated dataset.

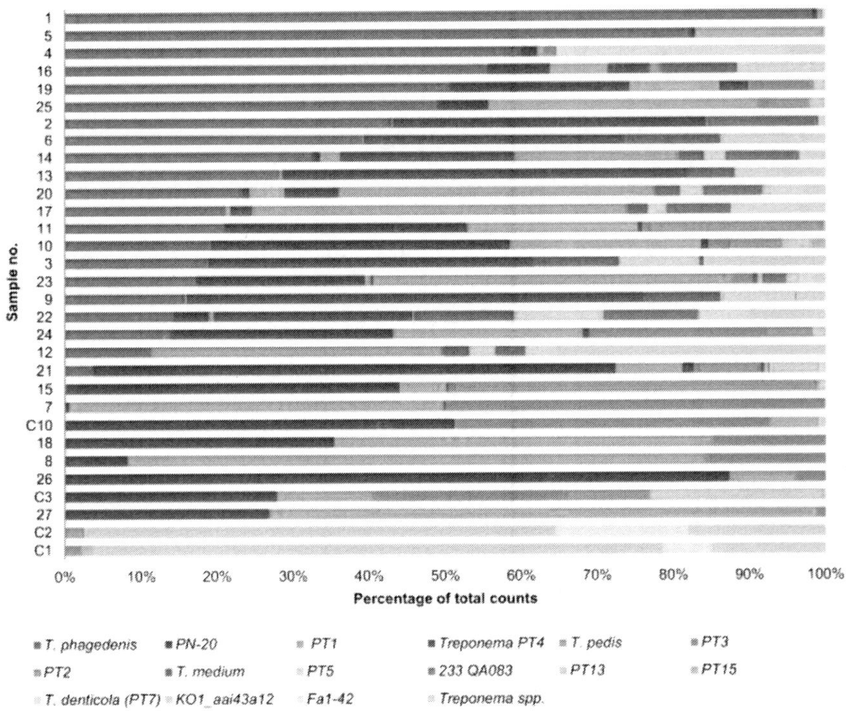

Figure 4. Relative abundance of *Treponema* phylotypes in the microbiota of bovine hoof skin samples sorted by the prevalence of *T. phagedenis*. Sequence amplicons were derived from 31 biopsy specimens by the use of *Treponema*-specific primers, V3–V4, and 16S rRNA gene-based metagenomics. C signifies samples with no visible lesions, the remaining samples all displayed macroscopic lesions.

3.6. In situ hybridization

Applying double in situ hybridization with the probes for Domain Bacterium and*Treponema* spp., severe and extensive treponemal epidermal infiltration, constituting more than 90% of the total bacterial population, was found in 24 of the 27 samples with scores 1 and 2 (Table 1, Fig. 1C). In two samples with ascore of 1 (no. 10 and 21), no treponemes were observed, and only a moderate number of bacteria were found infiltrating the epidermis. One sample (no. 18 with score 1) was omitted from in situ detection due to damage during sectioning. Invasive bacteria were not detected in any of the ten score 0 samples (only scarce bacteria on the epidermal surface); thus, further in situ detection was not carried out. *D. nodosus* was identified in 12 samples (4 score 1 cases and 8 score 2 cases) within mal-keratinized epidermis, in areas free of treponemes and other bacteria, as single bacteria or in loose clusters with up to 50 bacteria as illustrated in Fig. 1D. *F. necrophorum* was detected as single bacteria together with other bacteria (including treponemes) superficially in the epidermal lesions of eight samples (3 score 1 cases and 5 score 2 cases). In one case, however, *F. necrophorum* was detected in a cluster with more than 100 bacteria. Similarly, *P. levii* was detected as single bacteria or in small clusters, together with other bacteria superficially in the lesions, in eight score 2 samples as illustrated in Fig. 1B. The probes for *Mycoplasma* spp. and *M. fermentans*were tested on various type strains, including a *M. fermentans* strain isolated from humans, and they revealed moderate fluorescence signals. When applied on the case samples, no specific signal (positive both for Domain Bacterium and *Mycoplasma* spp./*M. fermentans*) could be identified.

3.7. RT-qPCR analysis of putative virulence genes

To determine the metabolic activity of the treponemes in the DD lesions, RT-qPCR was performed on 27 of the samples from which we could extract treponemal RNA (C2 and C3 plus all the DD-biopsies, except samples no. 8 and 18, which were omitted due to low concentration and quality of RNA). The target genes included in the analysis comprised putative virulence determinants and reference genes for normalization (Table 2). Because *T. denticola*, *T. vincentii* and *T. phagedenis* display considerable variation in their genomic sequences, it was not possible to design primers that could amplify all three species simultaneously for any of the selected target genes. Consequently, we constructed primers targeting each species and pooled the primers for the RT-qPCR. The specificity, dynamic range and amplification efficiency of the primers were tested on DNA extracted from *T. denticola* (ATCC33520), *T. vincentii* (ATCC35580) and T. *phagedenis*(ATCC27087) (Table S2 in the Supplemental Material). The relative quantity of each sample was estimated using an internal standard curve. The quantification values were normalized to the geometric mean of the two reference genes *tkt* and *tpiA*. All points outside the dynamic range were excluded from the analysis.

Of the 27 samples included, no signal or a very weak signal was generated from eight samples (samples no. C1, C2, 7, 9, 15, 22, 26 and 27). Thus, 19 samples remained in the analysis. In the case of the controls, the negative result was most likely due to the lack of target cDNA because most of the treponemes in these samples, according to the sequencing results, were environmental species. In samples no. 7, 15, 26 and 27, the overall abundance of treponemes was relatively low (Fig. 3). Furthermore, the*Treponema* amplicons present in these samples were mainly of the *T. refringens*-like phylotypes, which were not targeted in the RT-qPCR because no sequence was available in GenBank for primer design.

In the remaining 19 samples the *Treponema* target genes were successfully amplified (samples no. 1–5, 10–14, 16, 17, 19–21 and 23–25). The normalized expression values for the 19 samples are depicted in Fig. 5. The variability of template concentration was highest for *mglA* and *msp*, where approximately 1/3 of the samples had expression below 0.5 ng (the normalized template concentrations are shown in Table S3).

4. DISCUSSION

In this study we used high throughput sequencing, FISH and expression analysis to examine the microbial communities of normal and DD-afflicted bovine hoof skin. The 16S rRNA gene studies all revealed substantial differences in the bacterial microbiota composition between healthy samples and DD lesions. *Firmicutes*, *Actinobacteria*, *Proteobacteria*, and *Bacteroides*, the four main phyla identified in normal foot skin in this study, were also identified in healthy skin in at least one of the previous sequencing studies (Yano et al., 2010, Santos et al., 2012, Krull et al., 2014 and Zinicola et al., 2015). At the family level, *Moraxellaceae*, *Corynebacteriaceae*, *Lachnospiraceae* and *Rumino-coccaceae* were the most abundant in healthy samples.

Table 2. Genes targeted in the RT-qPCR analysis.

Gene name	GenBank locus no.	Possible function
Virulence genes		
Msp	TDE_0405	Major outer sheath protein antigen
FlaA	TDE_1408	Flagellar filament outer layer protein
FlgE	TDE_1007	Flagellar basal body rod protein
HbpB	TDE2055	Hemin-binding protein B, lipoprotein
MglA	TDE_2216	Galactose/methyl galactoside import ATP-binding protein
DnaK	TDE_0628	Chaperone protein (Hsp70)
Reference genes		
TpiA[a]	TDE_1236	Triosephosphate isomerase
Hex	TDE_2469	Hexokinase family protein
Tkt[a]	TDE_1308	Transketolase
Lnt	TDE_1708	Apolipoprotein N-acyltransferase

[a] Reference genes applied in the analysis for normalization.

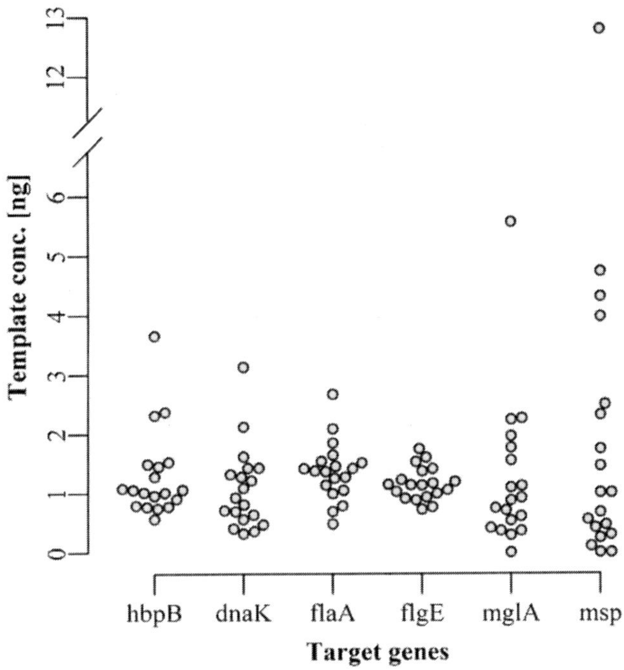

Figure 5. Scatter dot plot of normalized sample concentrations (n = 19) of *Treponema* target genes isolated from biopsies of DD lesions.

Most of the DD specimens in our investigation were from chronic lesions with areas of ulceration, corresponding to active lesions. We found *Spirochaetes* (47%) and*Tenericutes* (27.7%) to be the most ubiquitous bacterial phyla in the DD lesions, and very few *Firmicutes* (3.6%) were identified in these samples. So, contrarily to a number of previous studies (Santos et al., 2012 and Zinicola et al., 2015), we did not find evidence suggesting a pathogenic role for *Firmicutes* in DD. In this study, when probing the underlying microbial patterns associated with DD etiology, we found that *Treponema, Mycoplasma, Fusobacterium* and *Porphyromonas* were the genera best differentiating the infected tissue samples from controls. Our results are therefore mostly in accordance with the observations by Krull et al. (2014), who have also found the families*Spirochaetaceae* and *Mycoplasmataceae* to be the main constituents of clinical lesions.

Except for the work of Krull et al. (2014), who have hypothesized that *Mycoplasma* spp. may be of importance in the transition from early lesion to active lesion, *Mycoplasma* has hitherto not received much attention as a candidate pathogen—even though *M. fermentans* has previously been cultivated from the front of a DD lesion (Wyss et al., 2005) and identified by 16S rRNA gene sequencing (Santos et al., 2012 and Zinicola et al., 2015). *M. fermentans*, which is a member of the phylum *Tenericutes*, was originally isolated from the human urogenital tract and has since been associated with a number of human diseases (Pitcher and Hilbocus, 1998). In fact, it was first thought to be restricted to humans, but the species has since been isolated from genital lesions in sheep (Ayling et al., 2004). *M. fermentans*-specific PCR verified that this species was indeed present in the DD samples of the present study. However, the inconclusive results of the M*ycoplasma*-specific FISH analysis, combined with the negative results for IS*1550* and*malp* in the *M. fermentans*-specific RT-PCR, seem to imply that these bacteria are not very active in the lesions at the stages of the disease represented here.

Amplicons homologous to *F. necrophorum* and *P. levii* were the third and fourth most abundant variants in the DD biopsies, respectively, and these species have previously been identified in DD lesions (Cruz et al., 2005, Wyss et al., 2005 and Rasmussen et al., 2012). In this study, *Porhyromonas* was significantly associated with ulcerated lesions (stage 2). *F. necrophorum* and *P. levii* are mutually involved in the etiology of foot rot in ruminants (Walter and Morck, 2002 and Zhou et al., 2009). Furthermore, antibodies against *F. necrophorum* and *P. levii* have been detected in dairy cattle with DD (Moe et al., 2010). However, both *F. necrophorum* and *P. levii* displayed much lower abundance and prevalence than *Treponema* and *Mycoplasma* spp. (Fig. 3), and the FISH results showed them to be mainly located in the superficial layers of the infected epidermis. Consequently, their possible roles in the pathogenesis of DD remain unclear. They may be actively involved in some specific stages of the disease – or they may merely represent a secondary colonization as previously indicated (Rasmussen et al., 2012 and Krull et al., 2014). Another bacterium, *D. nodosus*, also suspected to be involved in the development of DD (Rasmussen et al., 2012) and recently shown to cause interdigital dermatitis in a challenge model (Knappe-Poindecker et al., 2015), was identified by FISH analysis in 12 of the samples but was absent or below the detection limit in the sequence data. Likewise, we did not observe *Candidatus Amoebophilus asiaticus* to be part of the microbiomes of the present DD samples. This bacterium is among the most abundant species observed in both active and inactive lesions in the study by Zinicola et al. (2015).

Although very few *Spirochaetes* were identified in the control samples, this family was amply represented in the DD lesions. This observation was further supported by the FISH analysis, which showed *Treponema* spp. to be the dominating phyla in the deeper parts of the lesions, constituting approximately 90% of the infecting microbiota. Most of the DD specimens in our investigation were active lesions (score 2) that exhibited changes in both the epidermis and dermis layer and disruption of the basement membrane. As previously observed by our group and others (Evans et al., 2009 and Klitgaard et al., 2013), there is a high diversity of *Treponema* species in the DD afflicted tissue, with *T. phagedenis* (including PT13) and *T. refringens*-like phylotypes (PT1-3, PT12 and PN-20) as the most abundant species (Fig. 4). Similar *Treponema* phylotypes have been detected in the deep sequencing studies by Krull et al. (2014) and Zinicola et al. (2015)both of which have reported temporal or stage changes in the treponemal composition of the lesions. Because most of the lesions in the present investigation were representative of variations over the same stage, we were not able to infer any predictions regarding microbial community changes over time. Moreover, the use of different lesion scoring systems in the metagenomics studies of DD conducted so far (Krull et al., 2014 and Zinicola et al., 2015) illustrates how difficult it is to classify the stages of this disease, which—in combination with variations in 16S rRNA gene sequence taxonomic classification methods—makes direct comparison between studies challenging with regard to lesion development over time.

To study the expression of selected potential *Treponema* virulence factors during infection, RT-qPCR was applied. In 17 of the samples, we detected expression of the major outer sheath protein antigen (*msp*), which is one of the principle virulence determinants of *T. denticola* (Fenno et al., 1997). Two genes encoding flagellar proteins were also actively expressed in 19 DD samples. Motility may be important for a bacterium's ability to actively invade host tissue (Dashper et al., 2011). To survive *in vivo*, pathogenic bacteria have developed several mechanisms for acquiring iron in the host environment. The protein encoded by *hbpB* may contribute to hemin utilization by *T. denticola* (Xu et al., 2001) and was actively expressed by treponemes in this study. Along with the FISH analysis, these results further indicate that treponemes are actively involved in the pathology of DD.

5. CONCLUSION

We observed a clear distinction between the microbiota of healthy and DD-afflicted tissue.*Treponema, Mycoplasma, Fusobacterium* and *Porphyromonas* were the genera best differentiating the infected tissue samples from the controls. FISH analysis indicated that species belonging to the genus *Treponema* were the dominant bacteria in the deep part of the lesions, and RT-qPCR indicated that they were actively expressing putative virulence genes. *F. necrophorum* and *P. levii* were mainly observed in the superficial parts of the epidermis by FISH. Although the presence of *M. fermentans* was confirmed by species-specific PCR, it could not be detected by FISH or by RT-PCR. Although the involvement of a variety of *Treponema* species in DD lesion development seems well supported, the roles of other bacterial genera are still unclear. For example, it seems puzzling that *M. fermentans* appeared to be highly abundant in the infected tissue, yet appeared to be relatively inactive. *F. necrophorum* and *P. levii* may be secondary invaders, or they could also play important roles at specific stages of infection. To further clarify this notion, metatranscriptomic studies of DD microbiota during infection are presently ongoing in our group.

CONFLICT OF INTEREST

None of the authors have any conflict of interest.

ACKNOWLEDGEMENTS

This work was supported by the Danish Research Council for Technology and Production Sciences (12–127537) and The Danish Dairy Levy Foundation (Mælkeafgiftsfonden). The authors thank the staff at Danish Crown in Aalborg for kind assistance and for providing sample material. We thank Annie Ravn Pedersen for excellent technical assistance.

REFERENCES

1. Afshar, B., Pitcher, D., Nicholas, R.A., Miles, R.J., 2008. An evaluation of PCR methods to detect strains of Mycoplasma fermentans. Biologicals 36, 117–121.

2. Anderson, M.J., Willis, T.J., 2003. Canonical analysis of principal coordinates: a useful method of constrained ordination for ecology. Ecology 84, 511–524.

3. Ayling, R.D., Bashiruddin, S.E., Nicholas, R.A., 2004. Mycoplasma species and related organisms isolated from ruminants in Britain between 1990 and 2000. Vet. Rec 155, 413–416.

4. Choi, B.K., Nattermann, H., Grund, S., Haider, W., Göbel, U.B., 1997. Spirochetes from digital dermatitis lesions in cattle are closely related to treponemes associated with human periodontitis. Int. J. Syst. Bacteriol. 47, 175–181.

5. Cruz, C.E., Pescador, C.A., Nakajima, Y., Driemeier, D., 2005. Immunopathological investigations on bovine digital epidermitis. Vet. Rec. 157, 834–840.

6. Döpfer, D., Koopmans, A., Meijer, F.A., Szakáll, I., Schukken, Y.H., Klee, W., Bosma, R. B., Cornelisse, J.L., van Asten, A.J., ter Huurne, A.A., 1997. Histological and bacteriological evaluation of digital dermatitis in cattle: with special reference to spirochaetes and Campylobacter faecalis. Vet. Rec. 140, 620–623.

7. Dashper, S.G., Seers, C.A., Tan, K.H., Reynolds, E.C., 2011. Virulence factors of the oral spirochete Treponema denticola. J. Dent. Res. 90, 691–703.

8. Demirkan, I., Walker, R.L., Murray, R.D., Blowey, R.W., Carter, S.D., 1999. Serological evidence of spirochaetal infections associated with digital dermatitis in dairy cattle. Vet. J. 157, 69–77.

9. Edwards, A.M., Dymock, D., Jenkinson, H.F., 2003. From tooth to hoof: treponemes in tissue-destructive diseases. J. Appl. Microbiol. 94, 767–780.

10. Evans, N.J., Brown, J.M., Demirkan, I., Singh, P., Getty, B., Timofte, D., Vink, W.D., Murray, R.D., Blowey, R.W., Birtles, R.J., Hart, C.A., Carter, S.D., 2009. Association of unique, isolated treponemes with bovine digital dermatitis lesions. J. Clin. Microbiol. 47, 689–696.

11. Fenno, J.C., Wong, G.W., Hannam, P.M., Muller, K.H., Leung, W.K., McBride, B.C.,1997. Conservation of msp, the gene encoding

the major outer membrane protein of oral Treponema spp. J. Bacteriol. 179, 1082–1089.

12. Hill, T.C., Walsh, K.A., Harris, J.A., Moffett, B.F., 2003. Using ecological diversity measures with bacterial communities. FEMS Microbiol. Ecol. 43, 1–11.

13. Klitgaard, K., Boye, M., Capion, N., Jensen, T.K., 2008. Evidence of multiple Treponema phylotypes involved in bovine digital dermatitis as shown by 16S rDNA analysis and fluorescent in situ hybridisation. J. Clin. Microbiol. 46, 3012–3020.

14. Klitgaard, K., Foix, B.A., Boye, M., Jensen, T.K., 2013. Targeting the treponemal microbiome of digital dermatitis infections by high-resolution phylogenetic analyses and comparison with fluorescent in situ hybridization. J. Clin. Microbiol. 51, 2212–2219.

15. Knappe-Poindecker, M., Jørgensen, H.J., Jensen, T.K., Tesfamichael, B., Ulvund, M.J., Hektoen, L., Fjeldaas, T., 2015. Experimental infection of cattle with ovine Dichelobacter nodosus isolates. Acta Vet. Scand. 57, 55.

16. Krull, A.C., Shearer, J.K., Gorden, P.J., Cooper, V.L., Phillips, G.J., Plummer, P.J., 2014. Deep sequencing analysis reveals temporal microbiota changes associated with development of bovine digital dermatitis. Infect. Immun. 82, 3359–3373.

17. Laven, R.A., Logue, D.N., 2006. Treatment strategies for digital dermatitis for the UK. Vet. J. 171, 79–88.

18. Losinger, W.C., 2006. Economic impacts of reduced milk production associated with papillomatous digital dermatitis in dairy cows in the USA. J. Dairy Res. 73, 244– 256.

19. Moe, K.K., Yano, T., Misumi, K., Kubota, C., Nibe, K., Yamazaki, W., Muguruma, M., Misawa, N., 2010. Detection of antibodies against Fusobacterium necrophorum and Porphyromonas levii-like species in dairy cattle with papillomatous digital dermatitis. Microbiol. Immunol 54, 338–346.

20. Moter, A., Leist, G., Rudolph, R., Schrank, K., Choi, B.K., Wagner, M., Göbel, U.B., 1998. Fluorescence in situ hybridization shows spatial distribution of as yet uncultured treponemes in biopsies from digital dermatitis lesions. Microbiology 144, 2459–2467.

21. Pitcher, D., Hilbocus, J., 1998. Variability in the distribution and composition of insertion sequence-like elements in strains of Mycoplasma fermentans. FEMS Microbiol. Lett 160, 101–109.

22. Pringle, M., Bergsten, C., Fernström, L.L., Höök, H., Johansson, K.E., 2008. Isolation and characterization of Treponema phagedenis-like spirochetes from digital dermatitis lesions in Swedish dairy cattle. Acta Vet. Scand 40 (50) .

23. Rasmussen, M., Capion, N., Klitgaard, K., Rogdo, T., Fjeldaas, T., Boye, M., Jensen, T.K., 2012. Bovine digital dermatitis: possible pathogenic consortium consisting of Dichelobacter nodosus and multiple Treponema species. Vet. Microbiol 160, 151– 161.

24. Rozen, S., Skaletsky, H., 2000. Primer3 on the WWW for general users and for biologist programmers. Methods Mol. Biol 132, 365–386.

25. Santos, T.M., Pereira, R.V., Caixeta, L.S., Guard, C.L., Bicalho, R.C., 2012. Microbial diversity in bovine papillomatous digital dermatitis in Holstein dairy cows from upstate New York. FEMS Microbiol. Ecol. 79, 518–529.

26. Vandesompele, J., De Preter, K., Pattyn, F., Poppe, B., Van Roy, N., De Paepe, A., Speleman, F., 2002. Accurate normalization of real-time quantitative RT-PCR data by geometric averaging of multiple internal control genes. Genome Biol. 3 (research0034).

27. Walter, M.R., Morck, D.W., 2002. Chemotaxis, phagocytosis, and oxidative metabolism in bovine macrophages exposed to a novel interdigital phlegmon (foot rot) lesion isolate. Porphyromonas levii. Am. J. Vet. Res. 63, 757–762.

28. Wang, R.Y., Hu, W.S., Dawson, M.S., Shih, J.W., Lo, S.C., 1992. Selective detection of Mycoplasma fermentans by polymerase chain reaction and by using a nucleotide sequence within the insertion sequence-like element. J. Clin. Microbiol. 30, 245–248.

29. Wilmotte, A., Van der Auwera, G., De Wachter, R., 1993. Structure of the 16S ribosomal RNA of the thermophilic cyanobacterium Chlorogloeopsis HTF ('Mastigocladus laminosus HTF') strain PCC7518 and phylogenetic analysis. FEBS Lett. 317, 96–100.

30. Wyss, C., Dewhirst, F.E., Paster, B.J., Thurnheer, T., Luginbuhl, A., 2005. Guggenheimella bovis gen. nov. sp. nov., isolated from lesions of bovine dermatitis digitalis. Int. J. Syst. Evol. Microbiol. 55, 667–671.

31. Xu, X., Holt, S.C., Kolodrubetz, D., 2001. Cloning and expression of two novel hemin binding protein genes from Treponema denticola. Infect. Immun 69, 4465–4472.

32. Yano, T., Moe, K.K., Yamazaki, K., Ooka, T., Hayashi, T., Misawa, N., 2010. Identification of candidate pathogens of papillomatous digital dermatitis in dairy cattle from quantitative 16S rRNA clonal analysis. Vet. Microbiol 143, 352–362.

33. Zhou, H., Bennett, G., Hickford, J.G., 2009. Variation in Fusobacterium necrophorum strains present on the hooves of footrot infected sheep goats and cattle. Vet. Microbiol. 135, 363–367.

34. Zinicola, M., Lima, F., Lima, S., Machado, V., Gomez, M., Döpfer, D., Guard, C., Bicalho, R., 2015. Altered microbiomes in bovine digital dermatitis lesions, and the gut as a pathogen reservoir. PLoS. ONE. 10, e0120504.

35. Zuerner, R.L., Heidari, M., Elliott, M.K., Alt, D.P., Neill, J.D., 2007. Papillomatous digital dermatitis spirochetes suppress the bovine macrophage innate immune response. Vet. Microbiol. 125, 256–264.

CHAPTER 7

Sampling and Pooling Methods for Capturing Herd Level Antibiotic Resistance in Swine Feces using qPCR and CFU Approaches

Gunilla Veslemøy Schmidt1*, Anders Mellerup1‡ , Lasse Engbo Christiansen2, Marie Ståhl1‡ , John Elmerdahl Olsen3, Øystein Angen1¤

1 The National Veterinary Institute, Technical University of Denmark, Frederiksberg C, Denmark,

2 Department of Applied Mathematics and Computer Science, Technical University of Denmark, Kongens Lyngby, Denmark,

3 Department of Veterinary Disease Biology, Faculty of Health and Medical Sciences, University of Copenhagen, Frederiksberg C, Denmark

ABSTRACT

The aim of this article was to define the sampling level and method combination that captures antibiotic resistance at pig herd level utilizing qPCR antibiotic resistance gene quantification and culture-based quantification of antibiotic resistant coliform indicator bacteria. Fourteen qPCR assays for commonly detected antibiotic resistance genes were developed, and used to quantify antibiotic resistance genes in total DNA from swine fecal samples that were obtained using different sampling and pooling methods. In parallel, the number of antibiotic resistant coliform indicator bacteria was determined in the same swine fecal samples. The results showed that the qPCR assays were capable of detecting differences in antibiotic resistance levels in individual animals that the coliform bacteria colony forming units (CFU) could not. Also, the qPCR assays more accurately quantified antibiotic resistance genes when comparing individual sampling and pooling methods. qPCR on pooled samples was found to be a good representative for the general resistance level in a pig herd compared to the coliform CFU counts. It had significantly reduced relative standard deviations compared to coliform CFU counts in the same samples, and therefore differences in antibiotic resistance levels between samples were more readily detected. To our knowledge, this is the first study to describe sampling and pooling methods for qPCR quantification of antibiotic resistance genes in total DNA extracted from swine feces.

1. INTRODUCTION

Antibiotic resistance in pathogenic bacteria is an increasing problem challenging disease treatment in humans and animals globally [1–3]. It is important to minimize antibiotic use in intensive agricultural practices where widespread antibiotic use is common e.g. in the pig production, as such use can have severe consequences for human health [4]. As resistance is selected for in both commensal and pathogenic bacteria during antibiotic treatment, it is of major concern if commensal bacteria become a reservoir of antibiotic resistance genes for pathogenic bacteria [1,4–6].

Surveillance of the presence of antibiotic resistance in individuals, populations and/or the environment facilitates risk management and may also support correct choice of drug for disease treatment. At present, no good strategy has been published for quantification of resistance levels in animals at herd level. When designing such a strategy, it is important to consider the combination of sampling and analytical methods in order to gain representation of the true resistance level, whilst a feasible time frame and economic resources must be maintained.

The traditional phenotypic methods for surveillance of antibiotic resistance in populations rely on cultures of indicator bacteria such as *Escherichia coli*, *Enterococcus* spp.,*Salmonella* spp., and *Campylobacter* spp. However, surveillance by this approach neglects the remaining intestinal microbiota and potentially underestimates the true antibiotic resistance levels in the bacterial community [7]. Therefore, the amount of resources put into the development of nucleic acid-based methods utilizing total bacterial community DNA for antibiotic resistance detection has increased vastly [7,8]. These methods enable detection and quantification of antibiotic resistance, also in the slow-growing, non-cultivable, viable, and non-viable bacteria. Multiple target genes of interest can easily be targeted within the community where bacteria could be sharing resistance determinants thus reflecting the entire gene pool [7]. Real-time PCR (qPCR) is an extremely sensitive method and has been the choice of method for specific and rapid quantification of genes, including genes encoding antibiotic resistance in microbiome samples such as fecal samples and fecal contaminated environments [7,9–11].

The aim of the study was to define the combination of sampling level and quantification method that captures antibiotic resistance at herd level. Herd level antibiotic resistance estimates by qPCR and resistance levels in coliform bacteria were used to analyze swine fecal samples collected at different levels (with and without pooling). The genes included in this study were chosen with disregard to the indicator bacteria as the point was to assess the two methods separately as two approaches to antibiotic resistance surveillance. We recommend either pen floor samples or shoe cover sampling for capturing herd level antibiotic resistance as both sampling types were able to quantify antibiotic resistance genes in swine feces.

2. MATERIALS AND METHODS

2.1 Sample collection

Pig fecal samples were collected from a single feeder pig operation in Denmark at 2 separate time points (sampling 1, during March 2013; sampling 2, during July 2013). The sampling methods were included for the following reasons: Shoe cover sampling is used in the Danish *Salmonella* control program in broiler flocks [12] and it is of interest to investigate whether shoe cover sampling can be applied to pig herds; pen floor sampling is currently used in the Danish *Salmonella* control program in pig herds [13,14]; slurry tank sampling is of interest as the antibiotic resistance levels within the slurry tank may reflect the herd's levels rendering it a convenient method for herd sampling potentially representing resistance levels for the past 6 months (they are emptied twice yearly). The collected samples, their sizes and how they were pooled prior to analysis are summarized in Table 1. Consent for fecal sample collection was given by the farmer. The sampling was performed by authorized Danish veterinarians and as the sampling did not include invasive handling of animals, no permit from the Animal Inspectorate was required.

In order to establish if pooled samples from the floor and shoe covers could be used instead of sampling individual animals, fecal samples were collected from a single stable (stable 1) during sampling 1 (Table 1). In the stable, 4 pens (pens 1–4) were randomly chosen, and within these 4 pens fecal samples from all pigs (pen 1 n = 22, pen 2 n = 19, pen 3 n = 21, pen 4 n = 19; total n = 84) were collected from the rectum. Furthermore, each of the 4 pens was sampled by pooling 5 separate samples of feces from the floor of each pen (pen floor samples, n = 4). Four shoe cover samples were also collected in the same 4 pens. Two people each wore a pair of blue disposable polypropylene shoe covers (SEPA, Hilleroed, Denmark) and the pen area was covered by walking throughout the entire pen in a systematic snake formed pattern thus covering as much of the pen floor as possible. The same shoe covers were repeatedly used in all 4 pens and were removed when walking between pens.

In order to establish if pooled samples from the floor, shoe cover samples, and slurry tank samples capture antibiotic resistance at herd level to the same degree, a total of 5 stables were sampled during sampling 2 (Table 1). Individual animal sampling was not included as this stage, since the relation between individual and pooled samples were determined above. During sampling 2, 4 pens were randomly chosen in each stable and sampled (n = 20). The pen floor samples and shoe cover samples were collected in each stable as described above with the exception that 2 shoe cover samples were collected for each stable (pen floor samples n = 20; shoe cover samples n = 10) instead of 4. Furthermore, slurry tank samples were taken at 1 m, 1.5 m, and 2 m depths from 3 sampling spots that were spaced approximately 50 cm apart. One ml from each of the 3 depths was mixed for each corresponding sampling spot and homogenized by vortexing. Each slurry tank sample thus consisted of a pool of the 3 depths (denoted as slurry tank 1, slurry tank 2, and slurry tank 3; total n = 3).

All samples were collected in plastic containers with tight lids or sealed plastic bags and were placed in coolers immediately after sampling. They were then stored at 5°C until analysis in the laboratory the following day. Thereafter, all samples were frozen at -80°C.

Table 1. Overview of the samples collected during sampling 1 and sampling 2 including different sampling methods and corresponding laboratory pools at stable and herd levels.

Pool name	Sampling 1 1 stable sampled	Sampling 2 5 stables sampled	Samples included in laboratory pool	Number of samples (n)
All animals	+	-	All individual animals from sampling 1 (extraction from rectum)	n = 84
Individual animal pool pen	+	-	Pool of individual animals within each pen (extraction from rectum)	Pen 1 n = 22;pen 2 n = 20;pen 3 n = 22;pen 4 n = 20
Pen floor samples	+	+	Not pooled	Sampling 1 n = 4; sampling 2 n = 20
Pen floor pool stable	+	+	Pool of pen floor samples from pens 1–4 in each stable	Sampling 1 n = 1;sampling 2 n = 5
Pen floor pool herd	-	+	Pool of pen floor samples from each stable (1–5)	n = 1
Shoe cover samples	+	+	Not pooled	Sampling 1 n = 4;sampling 2 n = 10
Shoe cover pool stable	+	+	Pool of the shoe cover samples in each stable	Sampling 1 n = 1;sampling 2 n = 2
Shoe cover pool herd	-	+	Pool of shoe cover samples from each stable (1–5)	n = 1
Slurry tank samples	-	+	Each sample was a pool from 3 depths collected at the same spot (1m, 1.5m, and 2m)	n = 3
Pool slurry	-	+	Pool of slurry tank samples 1–3	n = 1

2.2 Sample Processing and Pooling

The non-sampled shoe cover weight was subtracted from the sample shoe cover weights, and defined amounts of PBS buffer were added to the sampled shoe covers. This dilution was corrected for in the statistical calculations. The shoe cover samples were then vigorously vortexed for 1 minute in order to extract as much sample from the shoe cover as possible. Thereafter, each shoe cover was "wrung out" to extract the remaining liquid. A 10^{-1}dilution was made from the resulting solutions by mixing 1 ml of the shoe cover sample solution with 9 ml PBS buffer.

Pooled samples were created in the laboratory as outlined in Table 1 and consisted of: 1) Pool of all individual animals from sampling 1 (All Animals), 2) Pools of animals within each pen (individual animal pool pen), 3) Pool of pen floor samples from pens 1–4 in each stable (pen floor pool stable), 4) Pool of the shoe cover samples in each stable (shoe cover pool stable), 5) Pool of pen floor samples from each stable (pen floor pool herd), 6) Pool of shoe cover samples from each stable (shoe cover pool herd), 7) Slurry tank samples (pool slurry). Before pooling the samples in the laboratory, a 10^{-1} dilution of the sample was made by suspending 1 g of feces in 9 ml PBS buffer. The pool samples were made by mixing 1 ml of the 10^{-1} resolution for each corresponding sample included in the pool for 1 minute.

2.3 CFU counts of coliform bacteria

Coliform bacteria are often used as indicator bacteria in antibiotic resistance surveillance programs and were used in the current study [15,16]. The 6x6 drop plate procedure was used for CFU counts of coliform bacteria [17]. Briefly, ten-fold serial dilutions from each swine fecal sample collected were made. One drop (20 μl) of each dilution was carefully placed on MacConkey plates (Oxoid) without and with antibiotics (Ampicillin (16 mg l^{-1}); Erythromycin (32 mg l^{-1}); Sulfamethoxazole (256 mg l^{-1}); Tetracycline (16 mg l^{-1})) and incubated at 37°C for 24 hours.

2.4 DNA extraction for quantification by qPCR

DNA was extracted from the 10^{-1} dilutions of the swine fecal samples. Sample preparation consisted of homogenization of the 10^{-1} dilutions by vigorous vortexing with a 5 mm stainless steel bead (Qiagen, Copenhagen, Denmark). Thereafter, 350 µl of the 10^{-1}dilutions was transferred to a new eppendorf tube and put on ice. The samples were then lyzed for 1 minute (15 Hz at room temperature) in Tissuelyser II (Qiagen) followed by centrifugation for 90 seconds at 10000 rpm. The supernatant (300 µl) was transferred to a new eppendorf tube and 20 µl Proteinase K (20 mg/ml) (Promega, Roskilde, Denmark) was added (not on ice). The samples were then immediately loaded into the QiaSymphony robot using the QiaSymphony DSP Virus/Pathogen Mini Kit (Qiagen) according to the manufacturer's instructions. The final elution volume was 85 µl.

Both a negative and positive DNA extraction control was run in parallel with the samples during each DNA extraction. The negative extraction control was water and was a control to test for sample contamination during the extraction process. The positive extraction control was included to ensure that antibiotic resistance gene levels were not affected by the DNA extraction process, and consisted of a modified pig feces sample positive for the majority of antibiotic resistance genes included in this study. The genes that were not present in this sample, defined as having a cycle number (Cq)>30 in qPCR, were spiked into the sample by adding 100 µl 10^5 amplicons μl^{-1} to a final volume of 1500 µl extraction control. This corresponded to Cq values between 20 and 30.

2.5 Primers and probes

Fourteen antibiotic resistance gene determinants that have previously been associated with swine manure (*tet*(-A,-B,-C,-M,-O,-W) [9,18–22], *ermB/F* [10,11,22], *sulI/sulII* [23,24], *vanA*[22], *bla*$_{CTX\text{-}M\text{-}1}$ group [25], *bla*$_{CMY\text{-}2}$], and *bla*$_{SHV}$ family [11,26]) were selected. Other inclusion criteria for the antibiotic resistance genes included: The use of the antibiotic class in the Danish pig production, the association of the gene with PCR detection, and the occurrence of the gene in a wide bacterial population. It also had to be feasible to design a qPCR assay for the chosen genes utilizing the same temperature profile.

Sequences of the chosen antibiotic resistance determinant were retrieved from GenBank (during 2011–2012), and the sequences described in intestinal commensals and in pigs were chosen for alignment using ClustalX [27]. The conserved regions in the alignments were then used for primer and probe design using Primer3Plus Web Interface (Free Software Foundation, Boston, MA, USA). Previously published primers with annealing temperatures of 60–61°C and probes with annealing temperatures of approximately 70°C were used and modified if necessary (Table 2).

Potential primer and probe sequences were used to query GenBank DNA sequences using Basic Local Alignment Search Tool nucleotide (BLASTn) to determine specificity. The primer and probe sequences that matched the desired antibiotic resistant determinants were analyzed using Integrated DNA Technologies SciTools Oligoanalyzer (Integrated DNA Technologies, Inc., Coralville, IA, USA). Wherever necessary, degenerate bases were introduced into the primer/probe sequences to match all the sequences in the alignments. Primers (Table 2; S1 Table) were synthesized by TAG Copenhagen A/S (Frederiksberg, Denmark) and for 16S rDNA by DNA Technology A/S (Aarhus Denmark). Probes (Table 2) were synthesized by Applied Biosystems (Life Technologies, Naerum, Denmark).

The identity of all standard amplicons were verified by sequencing from both ends using the BigDye Terminator v3.1 Sequencing Kit on a 3130 Genetic sequencer (Applied Biosystems, Life Technologies). Due to the short amplicon lengths, sequencing attempts of the qPCR amplicon sequences failed. Therefore, the amplicon products were confirmed using the High Sensitivity DNA chip on an Agilent 2100 Bioanalyzer (Agilent Technologies, Walbronn, Germany). The qPCR assays were also tested for cross reaction by running each qPCR assay against a panel of 16 different antibiotic resistance gene amplicon negative controls at a concentration of $2x10^4$ copies μl^{-1}.

2.6 Generation of amplification standards

The standards used for absolute quantifications consisted of purified PCR amplicons from antibiotic resistance genes; this principle has successfully enabled absolute quantification of antibiotic resistance determinants previously [7,9,29]. The amplicon standards included in the present study were derived from bacterial strains or pig fecal samples (S2 Table). The bacterial strains positive for tetracycline and beta-lactam antibiotic resistance determinants were kindly provided by Yvonne Agersø (National Food Institute (DTU-FOOD), Lyngby, Denmark), those positive for sulphonamide resistance determinants and *ermB* by Anette M. Hammerum (Statens Serium Institut (SSI), Copenhagen, Denmark), the bacterial strain positive for *ermF* by Stefan Schwarz (Friedrich-Loeffler-Insitut (FLI) Neustadt-Mariensee, Germany), and the strain positive for *vanA* from Luca Guardabassi (Faculty of Medical and Health Sciences (SUND), University of Copenhagen). The fecal samples were provided by The Veterinary Institute, The Technical University (DTU-VET), Frederiksberg, Denmark.

Total DNA was extracted from the bacterial strains using Invitrogen-easy DNA kit (Invitrogen, Life Technologies, Naerum, Denmark) and from the fecal samples as described for the fecal samples collected for qPCR analysis. The amplicons were generated using the primers inS1 Table using a T3000 thermocycler for standard PCR (Biometra, Göttingen, Germany). The PCR Mastermix for *tet*(B), *tet*(C), *tet*(M), *tet*(O), *tet*(W), *sulI*, *sulII*, and *erm*B amplicon generation had the following concentrations per 25 µl reaction volume: 250 µM deoxynucleotide triphosphates (dNTPs), 1X buffer, 1.5 mM $MgCl_2$, 0.5 µM of each forward (FW) and reverse (RV) primers, 1.25 U Platinum *Taq* DNA Polymerase (Invitrogen, Life Technologies, Naerum, Denmark) plus 2 µl DNA. Cycling conditions were: Initial denaturation for 4 min at 95°, followed by 30 cycles PCR, each cycle consisting of 15 sec at 94°C, 30 sec at 58°C and 60 sec at 72°C. There was a final extension for 5 min at 72°C.

The *tet*(A) and *ermF* amplicons were generated as previously described (*tet*(A) [33], *ermF*[11] using the following reaction mix: 25 µl reaction volume containing 250 µM dNTPs, 1X buffer, 1.50 mM $MgCl_2$, 0.5 µM of each FW and RV primers, 1.25 U Platinum *Taq* DNA Polymerase (Applied Biosystems, Life Technologies) plus 2 µl DNA. $bla_{\text{CTX-M-1}}$ group,bla_{SHV} family, $bla_{\text{CMY-2}}$ and *vanA* amplicons were generated as previously described [26,34,35].

16S amplicons were generated as previously described [36] with the following modifications: The reaction mix contained 130 µg/ml of each FW and RV primer and 0.5 U Platinum *Taq* DNA Polymerase (Applied Biosystems, Life Technologies); the final elongation was extended to 5 min at 72°C.

Table 2. PCR primer and probe sequences.

Primers	Gene target	Sequence (5' 3')	Ann. temp (°C)	Amplicon size (bp)	GenBank access. no.[a]	Reference
FP_TETA_2	*tet*(A)	TTGGCATTCTGCATTCACTC	60	125 (840–974)	X00006	This study
RP_TETA_2		GAAGGCAAGCAGGATGTAGC	60			
PR_TETA_2		GATCACCGGCCCTGTAGCCG				
FP_TETB_Aminov[b]	*tet*(B)	TTACGTGAATTTATTGCTTCGG	60	206 (913–1119)	EF646764	[28] and this study
RP_TETB_Aminov		ATACAGCATCCAAAGCGCAC	60			
PR_TETB_Aminov_own		CGCCGACCAAATCGGTCAGA				
FP_TETC_6	*tet*(C)	GCCAGTCACTATGGCGTGCT	60	120 (124–244)	EU751613	This study
RP_TETC_6		CAAGTAGCGAAGCGAGCAGG	60			
PR_TETC_6		ACTGTCCGACCGCTTTGGCC				
FP_TETM_7	*tet*(M)	CAACGAGGACGGATAATACGC	60	191 (119–311)	X92947	This study
RP_TETM_7		CCATCTTTTGCAGAAATCAGTAGA	60			
PR_TETM_7		GGTGAACATCATAGACACGCCAGGA				
FP_TETO_Böck	*tet*(O)	AAGAAAACAGGAGATTCCAAAACG	60	75 (607–682)	AY660531	[29]
RP_TETO_Böck		CGAGTCCCCAGATTGTTTTTAGC	60			
PR_TETO_Böck		ACGTTATTTCCCGTTATCACGGAAGCG				
FP_TETW_Smith	*tet*(W)	GCAGAGCGTGGTTCAGTCT	60	66 (411–476)	AJ222769	[30]
RP_TETW_Smith		GACACCGTCTGCTTGATGATAAT	60			
PR_TETW_Smith		TTCGGGATAAGCTCTCCGCCGA				
FP_SUL1_2	*sulI*	ACGAGATTGTGCGGTTCTTC	60	159 (440–598)	EU056266	This study

RP_SUL1_2		CCGACTTCAGCTTTTGAAGG	60			
PR_SUL1_2		ACCGGCTCATCCTCGATCCG				
FP_SUL2_3	*sulII*	GATATTCGCGGTTTTCCAGA	60	141 (313–453)	AY360321	This study
RP_SUL2_3		CGCAATGTGATCCATGATGT	60			
PR_SUL2_3		AAGACGGGCAGGCAGATCGG				
FP_ERMB_Böck	*ermB*	GGATTCTACAAGCGTACCTTGGA	60	86 (390–476)	AB563188	[29]
RP_ERMB_Böck[b]		TGGCAGCTTAAGCAATTGCT	60			
PR_ERMB_Böck		CACTAGGGTTGCTCTTGCACACTCAAGTC				
FP_ERMF_KNAPP	*ermF*	TCGTTTTACGGGTCAGCACTT	60	182 (24–205)	M14730;M17124; M17808;M62487	[11] and this study
RP_ERMF_KNAPP		CAACCAAAGCTGTGTCGTTT	60			
PR_ERMF_OWN		ATATTGGGGCAGGCAAGGGGTT				
FP_vanA_Böck	*vanA*	CTGTGAGGTCGGTTGTGCG	60	64 (614–705)	AF516335	[29]
RP_vanA_Böck		TTTGGTCCACCTCGCCA	60			
PR_vanA_Böck		CAACTAACGCGGCACTGTTTCCCAAT				
FW3_SHV_lahey	*bla*$_{SHV}$ family	GCTGGAGCGAAAGATCCACT	60	247 (258–504)	All available at http://www.lahey.org/Studies/	[31] and this study
RV5_SHV_lahey		CGCCTCATTCAGTTCCGTTT	60			
Pr_SHV_Lahey2		AYGTCACCCGCCTTGACCGC				
FW3_CMY-2_Lahey	*bla*$_{CMY-2}$	AGACGTTTAACGGCGTGTTG	60	127 (260–387)	All available at http://www.lahey.org/Studies/	[31] and this study

RV4_CMY-2_Lahey		TAAGTGCAGCAGGCGGATAC	60			
PR_CMY-2_Lahey		TATCGCCCGCGGCGAAAT				
FW_CTX-M-1	*bla*$_{CTX-M-1group}$	ATGTGCAGYACCAGTAARGTKATGGC	58	335	X92506	[32]
RV_CTX-M-1		ATCACKCGGRTCGCCXGGRAT	58			
PR_CTX-M-1		CCCGACAGCTGGGAGACGAAACGT				
FW_SMI_114	16S rDNA	CGCGAAGAACCTTACC	60	126 (916–1041)	NA	The Public Health Agency of Sweden
R_SMI_115		ACTTAACCCAACATTTCAC	60			
PR_SMI_116		CACGAGCTGACGACAGCC				

Forward primer = FP; Reverse primer = RP; Probe = PR, gene targets, annealing temperatures (Ann. temp.), amplicon lengths in base pairs (bp), and GenBank sequence accession number (GenBank access. no.).

[a] GenBank accession numbers for previously published primers and/or probes. For primers and probes designed in this study, a GenBank accession number representative of those included in the sequence alignments is given.

[b] Primer modified to fit assay in the present study.

Amplicon lengths were confirmed by gel electrophoresis and gene copy numbers were calculated after DNA quantification by UV spectrophotometry using a NanoDrop 3300 (Thermo Scientific, Wilmington, DE, USA). The qPCR standards were created by serially diluting the target gene amplicons in nuclease-free yeast tRNA (1:100 tRNA dilutions of 10 mg mL^{-1} (Applied Biosystems, Life Technologies)). The limit of quantification (LOQ) was defined as the lowest point in the amplicon standard serial dilution where all triplicates were positive. The limit of detection (LOD) was defined as the lowest concentration in the amplicon standard serial dilution where at least 1 of the triplicates was positive.

2.7 Internal amplification control for qPCR assays

Internal control amplicons (ICA) consisting of lambda (λ) phage DNA flanked by the forward (FP) and reverse (RP) primer sequences of the respective antibiotic resistance genes were included in the respective qPCR assay as an internal amplification control (IC). The primers used were the antibiotic resistance qPCR assay primers (Table 2) with a λ phage DNA sequence added to the 3´end (extra sequences 5'-3'direction: FP ATGAATATGACCAGCCAAC, RP TTCACGCAGGGGAAAT ATCTTTC) [37]. The ICAs were generated on a T3000 thermocycler (Biometra, Göttingen, Germany) in a reaction volume of 50 µl with 50 µM $MgCl_2$, 1X Buffer, 10 µM dNTPs, 50 µM of each forward and reverse primers, 1.25 U Platinum *Taq* DNA Polymerase (Applied Biosystems, Life Technologies), and 1 µL λ DNA (1 ng μl^{-1}) (Applied Biosystems, Life Technologies). The cycling conditions were: 5 min incubation period at 94°C followed by 10X touchdown cycles from 58°C, each touchdown cycle consisting of 1 min at 94°C, 1 min at 58°C, and 1½ minute at 72°C. Thereafter, there were 5 cycles, each with 1 min at 94°C, 1 min at 48°C and 1½ min at 72°C followed by 8X touchdown cycling from 48°C, each touchdown cycle consisting of 1 min at 94°C, 1 min at 48°C and 1½ min at 72°C. Next there were 12 cycles, each with 1 min at 94°C, 1 min at 40°C, and 1½ min at 72°C. Finally, there was an elongation of 10 min at 72°C.

The ICA length of 690 bp was verified by gel electrophoresis and the ICA was serially diluted to 10^{-11}. All antibiotic resistance gene qPCR assays were run on Rotorgene thermocyclers (Rotorgene Q-5plex and Rotorgene Q (Qiagen, Copenhagen, Denmark) with a Coxlam probe (VIC label 5'- CCACGAAGCCGCACFACTCCGC; Applied

Biosystems, Life Technologies) and ICA added to the mastermix. In order to determine the concentration of ICA to use as an internal control the following was tested: 5 separate mastermixes with ICA PCR product dilutions within 10^{-3} to 10^{-11} and 1 “no internal control” were tested against the lowest 3 concentrations detectable by the respective antibiotic resistance gene qPCR assay. The ICA dilutions that did not inhibit the respective antibiotic resistance gene qPCR assays were used as the internal controls, where the λ DNA easy probe was detected on the yellow channel (530–555 nm) and the antibiotic resistance gene qPCR probes on the green channel (470–510 nm).

2.8 Quantification of antibiotic resistance genes in pig fecal samples by qPCR

Quantitative PCR amplifications for the quantification of *tet*(A),*tet*(B), *tet*(C), *tet*(M), *tet*(O),*tet*(W), *ermF*, *ermB*, *sulI*, *sulII*, $bla_{CTX\text{-}M\text{-}1}$ group, $bla_{CMY\text{-}2}$, bla_{SHV} family, *vanA* and 16S in total DNA extracted from pig fecal samples were performed with Rotorgene thermocyclers (Rotorgene Q-5plex and Rotorgene Q, 72-well rotor 1–72) (Qiagen). The mastermixes are depicted in Table 3 and cycling conditions were: 10 min incubation period at 95°C followed by 45 cycles of PCR, each cycle consisting of 15 sec at 94°C and 30 sec at 60°C with a single fluorescence reading at green and yellow channels at the end of the extension stage. Each sample was tested in duplicate, along with a single point from the tenfold dilution series of the specific standard in triplicate, a single negative template control (NTC) that was 23μl mastermix and 2 μl water, and 1 positive and negative DNA extraction control. Quantification was performed using standard curves obtained from the PCR generated positive controls.

2.9 The impact of pig fecal environment on quantification

The impact of pig fecal environment DNA on the quantification of the respective antibiotic genes, using the 14 antibiotic resistance gene qPCR assays and the 16S rDNA qPCR assay, was tested. Nuclease-free yeast tRNA amplicon serial dilutions were analyzed parallel to pig fecal DNA spiked with antibiotic resistance gene amplicon serial dilutions. Using serial dilutions of amplicons assures confidence of the number of templates that are added to start with when preparing the serial dilutions and assumptions of gene copy number per positive control bacteria are avoided, which may introduce bias if the bacteria lose their plasmid, have different plasmid copy numbers, or shed the antibiotic resistance gene during the extraction process [7].

2.10 Statistical analysis

All figures and statistical tests were completed using R software (Version 3.0.1). The geometric mean of the technical replicates of gene copy numbers were used for further analysis. When only one of the technical replicates was above the LOQ than that value was used. Differences in gene copy numbers g^{-1} feces and in coliform bacteria CFU counts between pens from sampling 1 were calculated using Kruskal-Wallis rank sum test.

2.11 Dataset information

The dataset is located at The National Veterinary Institute, Section for Bacteriology, Pathology and Parasitology, The Technical University of Denmark (latitude: 55° 41’ 4.34” N; longitude: 12° 32’46.65” E; elevation: 7.34 m).

3. RESULTS

3.1 Accuracy of the qPCR assays

Standard curves for qPCR were generated using the serial dilutions of the amplification standards. The dynamic ranges of the antibiotic resistance gene assays were all linear over a measurement range >7 orders of magnitude and 5 orders of magnitude for the 16S rDNA assay (S3 Table). The amplicon standard serial dilutions were used for determining the linear dynamic range where $R^2 = 0.99$, efficiency = [0.90; 1.10] and $M \approx -3.2$. The efficiencies of the qPCR assays, determination coefficient (R^2), dynamic range, quantification—and detection limits are all summarized in S3 Table.

Table 3. Concentrations of reagents per reaction used in qPCR assays with a total reaction volume of 25 µl including 2 µl DNA.

Gene	qPCR Mastermix					Forward Primer	Reverse primer	Probe	Λ Probe	Λ PCR pro-duct
	Taqman Universal (Applied Biosystems)	Buffer	$MgCl_2$mM	Platinum *Taq* Polymerase (Invitrogen, Life Technologies, Grand Island, NY, United States) U	dNTPs µM	µM	µM	µM	µM	ul
tet(A)	1X	-	-	-	-	0.5	0.5	0.2	0.2	1
tet(B)	_	1X	3.5	1.25	250	0.5	0.5	0.2	0.2	1
tet(C)	1X	-	-	-	-	0.5	0.5	0.2	0.2	1
tet(M)	-	1X	2.5	1.25	250	0.8	0.8	0.2	0.2	1
tet(O)	1X	-	-	-	-	0.6	0.6	0.2	0.2	1
tet(W)	1X	-	-	-	-	0.9	0.9	0.2	0.2	1
ermB	1X	-	-	-	-	0.5	0.5	0.2	0.2	1
ermF	-	1X	1.5	1.25	250	0.5	0.5	0.2	0.2	1
sulI	-	1X	1.5	1.25	250	0.8	0.8	0.2	0.2	1
sulII	1X	-	-	-	-	0.8	0.8	0.2	0.2	1
vanA	1X	-	-	-	-	0.6	0.6	0.2	0.2	1
$bla_{CTX-M-1}$ group	-	1X	1.5	1.25	250	0.5	0.5	0.2	0.2	1
bla_{CMY-2}	-	1X	3.0	1.25	250	0.6	0.6	0.4	0.2	1
bla_{SHV} family	-	1X	3.5	1.25	250	0.6	0.6	0.2	0.2	1
16S rDNA	1X	-	-	-	-	0.9	0.9	0.2	NA	NA

The qPCR primers did not give unspecific reactions when tested against a panel of 16 different antibiotic resistance gene amplicon negative controls at $2x10^4$ copies μl^{-1}. The amplicons of all specific qPCRs from fecal samples produced the expected specific gene sequence, without any non-specific amplification (data not shown).

The performance of the qPCR tests was further evaluated using extracted DNA from pig feces for spiking with antibiotic resistance genes. The efficiency remained between [0.90;1.10] and R^2 above 0.99 for all assays except for *tet*(A) which showed an efficiency of 0.85, *tet*(M) with an efficiency of 0.84 and R^2 = 0.87, *sulII* with an efficiency of 0.88, *vanA* with an efficiency of 0.87, and 16S rDNA with an efficiency of 0.84. Varying degrees of inhibition were observed when spiking the amplicons in pig fecal DNA extracts (1 to 4 Cq value increase), indicating a slight assay specific inhibition (data not shown). However, the dynamic ranges remained linear over a measurement range >4 orders magnitude in the spiked pig feces DNA environment (data not shown).

3.2 Antibiotic resistant coliform CFU counts and qPCR gene copy number assessment in individual and pooled samples

CFU counts of coliform bacteria were chosen as an approach to antibiotic resistance estimation in the collected swine fecal samples. Fig 1 (top) shows a boxplot over the CFU counts of resistant coliform bacteria in the individual animals within the 4 pens from sampling 1. There was a significant difference between pens for the ampicillin resistant CFUs only ($p<0.05$; data not shown).

Next, the pooled samples of individual animals within pens (Fig 1, top, solid circles) were evaluated in relation to the antibiotic resistance levels in the non-pooled individual animal samples The individual animal pool pen samples were largely dispersed compared to the median of corresponding non-pooled individual animal samples. This was most prominent in the erythromycin, sulphonamide, and tetracycline groups (Fig 1, top).

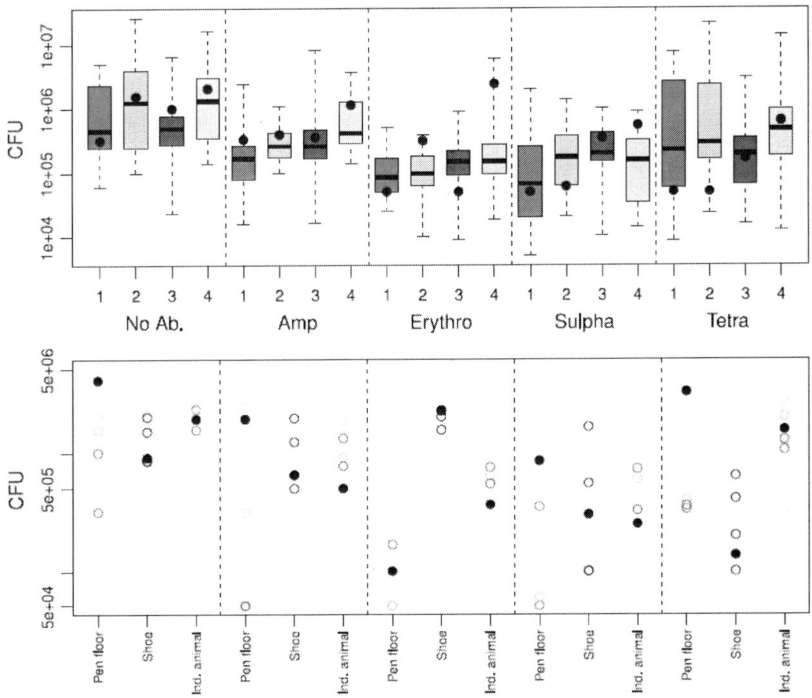

Figure 1. Sampling 1 coliform bacteria CFUs g^{-1} feces in individual animals within pens 1–4. Top: Boxplot showing the distribution of the coliform bacteria CFUs from sampling 1 in individual animals within pens 1–4 (pen 1, red, n = 22; pen 2, green, n = 20; pen 3, purple, n = 22; pen 4, blue, n = 20) on MacConkey plates (No Ab. = no antibiotics, Amp = ampicillin, Erythro = erythromycin, Sulpha = sulphonamide, Tetra = tetracycline). The bottom and top of the boxes are the first and third quartiles, respectively. The black band inside the box is the median and the "dotted-whiskers" represent the maximum (greatest CFU number values, excluding outliers) and minimum (lowest CFU value, excluding outliers). The solid circles are the individual animal pool pen samples within each pen. Bottom: The distribution of coliform bacteria CFUs from different sampling and pooling methods from sampling 1. The sampling methods are given under the bottom (Pen floor = pen floor sample; Shoe = shoe cover sample; Ind. animal = Individual animal sample). Each pen is represented by their colored circle (pen 1, red circle; pen 2, green circle; pen 3, purple circle; pen 4, blue circle), and the corresponding stable pools are solid black circles. The individual shoe cover samples are empty black circles, and the corresponding stable pool is a solid black circle.

Fig 1 (bottom) depicts the pen floor, shoe cover, and individual animal coliform CFU counts for each pen from sampling 1. The respective stable pools are also included (pen floor pool stable, shoe cover pool stable, and individual animal pool stable). There was a large variation within each sampling method with the pen floor samples having the largest range. The shoe cover and individual animal stable pools either lie among or below their corresponding pen pools, where the pen floor CFU counts were below the pen floor pool stable for all groups except ampicillin and erythromycin.

The *bla*$_{\text{CTX-M-1}}$ group, *bla*$_{\text{CMY-2}}$, *bla*$_{\text{SHV}}$, and *vanA* antibiotic resistance genes were not detected in any samples and were therefore excluded from further analysis and all graphs. The tendencies described for the gene copy estimates in sampling 1 and 2 did not change after normalization with 16S rDNA (S1 and S2 Figs). Therefore, only the absolute quantifications by qPCR were used for further data analysis.

Fig 2 (top) illustrates the copy number distribution of each gene for all animals sampled within each pen. Genes such as *sulII*, *ermB*, and *tet*(M) were relatively constant between pens while the genes *ermF*, *tet*(A), *tet*(C), *tet*(O), and *tet*(W) varied in at least 1 of the 4 sampled pens. Generally *ermB*, *ermF*, *tet*(O), and *tet*(W) had higher copy numbers g^{-1} feces compared to *sulI*, *sulII*, *tet*(A)-*tet*(C), and *tet*(M). *sulI* gene copy estimates were particularly low while *tet*(B) had levels below the LOQ in pens 2 and 3.

The gene copy numbers of *ermB*, *ermF*, *tet*(C), *tet*(O), and *tet*(W) were significantly different between pens ($p<0.05$ for *ermB*, *tet*(C), *tet*(W); $p<0.0001$ for *ermF* and *tet*(O)). Pen 1 consistently had lower gene copy number g^{-1} feces for *ermB*, *ermF*, *tet*(O), and *tet*(W) compared to pens 2–4 with *tet*(C) also having lower gene copy number g^{-1} feces in pen 2 (Fig 2 (top)). *ermF* in particular varied between pens with a large variation within pen 2.*tet*(A) generally had large variations within pens compared to the other genes with the highest levels in pens 1 and 3.

Fig 3 illustrates the relative standard deviations of the fecal estimates of the coliform CFU counts and qPCR gene copy numbers. The qPCR gene copy numbers have lower relative standard deviations compared to the coliform CFU counts. The estimated relative standard deviation was under 20% for only 50% of the coliform CFU counts, while the relative standard deviation of qPCR gene copy numbers was under 20% for 90% of the cases. Therefore, only qPCR gene copy number g^{-1} feces estimates were used to assess the different sampling and pooling methods.

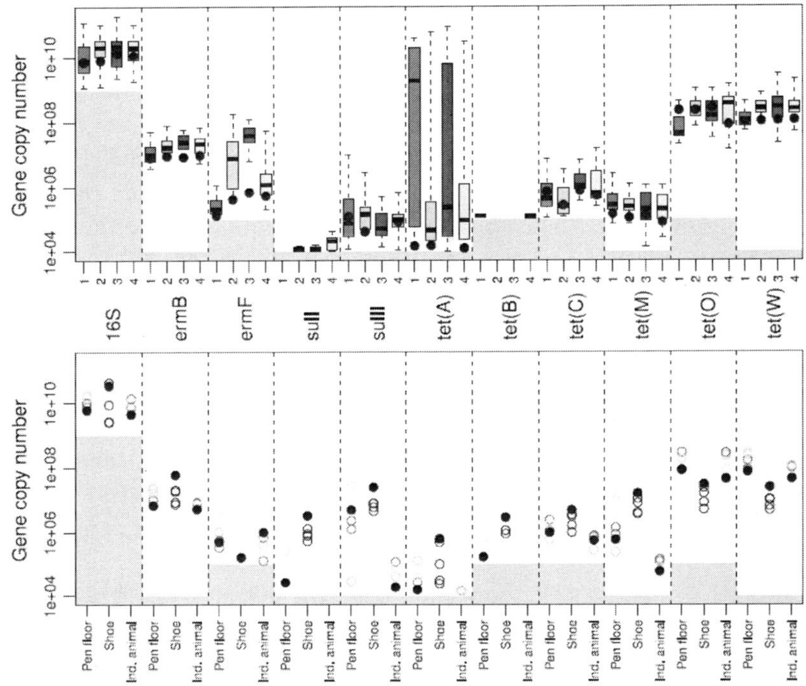

Figure 2. Sampling 1 gene copy numbers g^{-1} feces in individual animals within pens 1–4. Top: Boxplot showing the distribution of gene copies above the limit of quantification, LOQ (grey area = below LOQ) for individual animals within pens 1–4 from sampling 1 (pen 1, red, n = 22; pen 2, green, n = 20; pen 3, purple, n = 22; pen 4, blue, n = 20). The bottom and top of the boxes are the first and third quartiles, respectively. The black band inside the box is the median and the "dotted-whiskers" represent the maximum (greatest gene copy number values, excluding outliers) and minimum (least gene copy number value, excluding outliers). The solid circles are the individual animal pool pen samples within each pen. Each column represents a denoted gene (the respective genes are depicted in the middle of the figure and are shared for the top and bottom section of Fig 2). Bottom: The distribution of gene copies above the LOQ (grey area = below LOQ) for different sampling and pooling methods from sampling 1. The sampling methods are given under the bottom figure (Pen floor = pen floor sample; Shoe = shoe cover sample; Ind. animal = Individual animal sample). Each pen is represented by their colored circle for the pen floor and Ind. animal samples (pen 1, red circle; pen 2, green circle; pen 3, purple circle; pen 4, blue circle). The shoe cover samples are the same 4 shoe covers that were used in all 4 pens (individual shoe cover samples are empty black circles). The stable pools of the respective sampling method, pen floor, shoe cover, Ind. animal, are solid black circles.

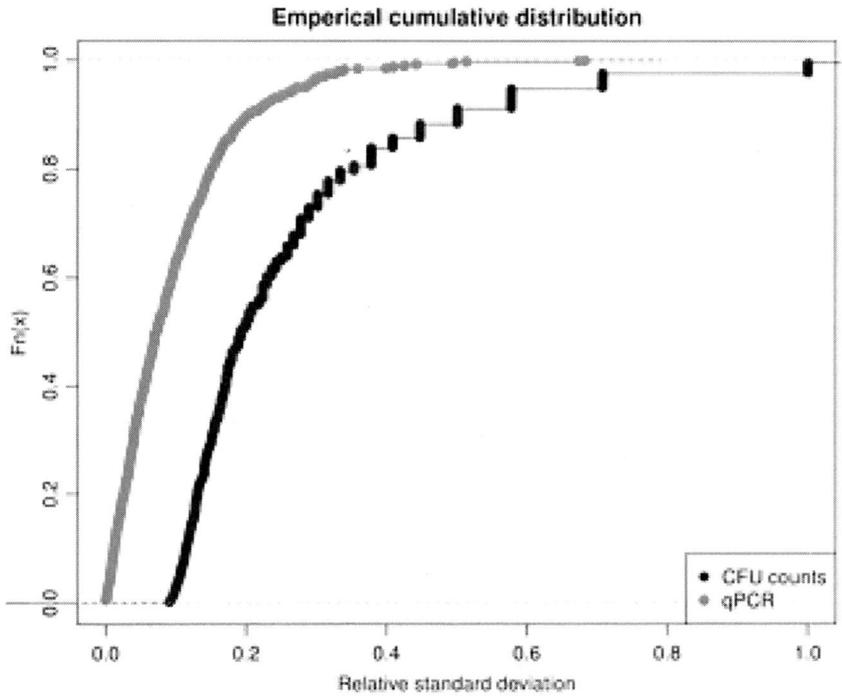

Figure 3. Relative standard deviations of the coliform CFU and qPCR gene copy number estimates g^{-1} feces. Empirical cumulative distribution (Fn(x)) plotted against the relative standard deviations of the CFU and gene copy number estimates illustrating the relationship between uncertainties of calculated estimates and the true laboratory determined estimates for CFU counts (black) and qPCR gene copy numbers (red), respectively.

The next step was to establish whether pooled samples from the floor and shoe covers cold be used instead of sampling individual animals. Fig 2 (bottom) depicts the distribution of each gene within pens 1–4 for the pen floor samples, shoe cover samples, pen floor pool stable, shoe cover pool stable, individual animal pool pen, and all animals. For all genes except for *ermF* and *tet*(C), there was a tendency for lower gene copy number estimates in the laboratory pools of individual animals within each pen when comparing to the pen floor samples. The shoe cover stable pools were consistently higher than the non-pooled shoe cover samples. In contrast, there were no *ermF* individual shoe cover samples above the LOQ but there were positive individual animal samples in all four pens (Fig 2 bottom), however the shoe cover pool stable was positive.

3.3 Pooling strategies at herd level

The final step was to establish if pooled samples from the floor, shoe cover samples, and slurry tank samples captured antibiotic resistance at herd level to the same degree. A comparison of results from sampling 2 including pen floor samples and shoe cover samples is shown in Fig 4 (top) together with the pooled samples from each category. Each section had up to 4 pen floor samples (1 from each pen) and 2 shoe cover samples with their respective laboratory pools. For *ermB*, *ermF*, *tet*(C), *tet*(O), and *tet*(W) the shoe cover samples gave lower estimates compared to the pen floor samples with the pools following the same pattern. For *sulI*, *sulII*, *tet*(A), *tet*(B), and *tet*(M), however, the shoe cover samples were higher than the pen floor samples. The 16S rDNA levels were stable regardless of the sampling and/or pooling methods for sampling 1 and sampling 2 (Figs 2 and 4).

Fig 4 (bottom) compares the pen floor and shoe cover pools both at stable and herd levels. The pooled and individual slurry samples from sampling 2 are also depicted in Fig 4(bottom). The assays that had shoe cover samples with higher levels than the pen floor samples within stables 1–5 from sampling 2 (*sulI*, *sulII*, *tet*(A), and *tet*(M) (Fig 4 top)) also had higher shoe cover samples in the corresponding stable and herd level pools (Fig 4bottom). The slurry tank samples complemented the pen floor and shoe cover herd pools, although it appeared that 1 of the 3 slurry tank samples was consistently lower than the others for all genes. Furthermore, *sulII* became negative at the pen floor herd pool despite having positive values in 2 out of 5 stable pools, and *tet*(A) had negative pen floor herd pools for and shoe cover herd pools despite their positive stable pools. *tet*(B) had a single positive shoe cover sample in stable 2 (Fig 4 top) with no positive results for stable pools, herd pools, and slurry tank samples (Fig 4 bottom). *tet*(C) had positive pen floor samples and pen floor pool stable samples in all stables except stable 4 (Fig 4 top). The *tet*(C) pen floor pool herd sample was positive (Fig 4 bottom), but the shoe cover and slurry tank samples were all negative (Fig 4 top and bottom).

4. DISCUSSION

Monitoring the antibiotic resistance patterns of infectious bacteria and their distribution aids disease prevention and control. Quantification of antibiotic resistance levels facilitates antibiotic resistance surveillance, ultimately helping to contain and prevent infections caused by antibiotic resistant bacteria. In the present study, 14 qPCR assays quantifying antibiotic resistant determinants and coliform bacteria CFU counts were used as two different approaches to antibiotic resistance estimation in swine fecal samples. This was done by applying both methods to swine fecal samples collected using different sampling and pooling methods. The main findings of our research were that the qPCR method detected significant differences in antibiotic resistance where the coliform CFU counts showed no significance, and furthermore, qPCR gene copy estimates in swine feces had reduced relative standard deviations compared to coliform CFU counts in the same samples, and therefore differences in antibiotic resistance levels between samples were more readily detected. This implicates that qPCR was a good representative for the general resistance level.

4.1 Accuracy of the qPCR assays

Primer design, DNA quality, thermal profile, and mastermix all play an important role in the specificity and quality of the PCR product, and sensitivity of the assay [38]. qPCR optimization is therefore required for each untested assay, with subsequent verification of assay specificity. In the current study, there was no cross reaction when each assay was tested against a panel of 16 different antibiotic resistance gene amplicons negative controls. Furthermore, the amplicons of all specific qPCRs from fecal samples produced the expected specific gene sequence, without any non-specific amplification (data not shown). The design of qPCR assays containing internal, positive- and negative-extraction controls ensured a thorough PCR inhibition and DNA extraction procedure control.

Assay inhibition in the spiked pig fecal DNA samples compared to water was seen as slightly lower efficiencies and higher corresponding Cq values in the spiked fecal DNA samples. This discrepancy between DNA from complex environmental samples has been described and was expected [39,40]. The dynamic range remained linear with a minimum 4-fold magnitude in the spiked pig fecal DNA samples for all assays (data not shown) which is sufficient for genomic DNA, and R^2 remained above 0.99 in all but 1 assay. Therefore, the slight variation seen in the spiked fecal DNA samples compared to sterile water is not considered to notably alter the assays' performance in quantifying antibiotic resistance genes in pig fecal samples. We believe that the DNA extraction protocol, primer sets, and corresponding probes possess the characteristics necessary for application to antibiotic resistance gene quantification in pig fecal samples.

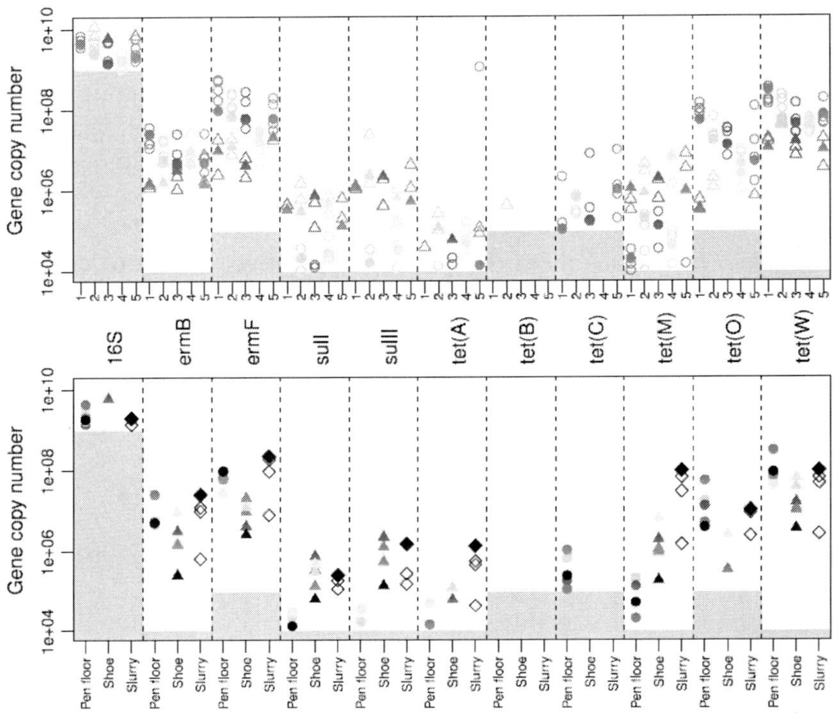

Figure 4. Sampling 2 gene copy numbers g^{-1} **feces in sections 1–5.** Top: Copy numbers of genes g^{-1} feces above the limit of quantification, LOQ (grey area = below LOQ) for pen floor samples (circles) and shoe cover samples (triangles) within sections 1–5 in sampling 2 (section 1, red; section 2, green; section 3, purple; section 4, blue; section 5, pink). The laboratory pooled samples are included for each section in their respective color (pen floor pool stable, solid circle; shoe cover pool stable, solid triangles). Each column represents a denoted gene (the respective genes are depicted in the middle of the figure and are shared for the top and bottom section of Fig 4). Bottom: Copy numbers of genes g^{-1} feces above the LOQ (LOQ = grey area) for pen floor pool stable samples (solid circles); shoe cover pool stable samples (solid triangles) for stables 1–5 (stable 1, red; stable 2, green; stable 3, purple; stable 4, blue; stable 5, pink); Slurry samples 1–3 (white diamond); Pen floor pool herd samples (black solid circle); Shoe cover pool herd sample (black solid triangle); Pool slurry (black solid diamond). The sampling methods are given under the bottom figure (Pen floor = pen floor pool herd; Shoe = shoe cover pool herd; Slurry sample).

4.2 Antibiotic resistant coliform CFU counts and qPCR gene copy number assessment in individual and pooled samples

When observing the coliform CFU counts in individual animals between pens, only a single group (ampicillin) showed a significant difference in resistant coliform bacteria. In contrast, significance was found in qPCR gene copy estimates for erythromycin and tetracycline, while no beta-lactamase genes were detected emphasizing how the two methods used for antibiotic resistance estimation yield each their endpoint. The phenotypic ampicillin resistance could be due to other ampicillin resistance encoding genes than those included in the present study emphasizing a key limit to the qPCR method, namely not all antibiotic resistance genes are included. In contrast, the phenotypic CFU counts are limited to coliforms representing only a fraction of the intestinal bacterial population, and may lead to underestimates of the true antibiotic resistance levels due to the limitations of the chosen indicator bacteria. Therefore, when monitoring antibiotic resistance, we argue that it is favorable to quantify the genes by qPCR instead of relying on phenotypic determination. This ensures that the entire bacterial population is represented while the denoted genes of interest (and therefore resistance) are also included. This principle is illustrated when assessing the coliform bacterial CFU counts in individual animals (Fig 1 top). Here, the erythromycin group had some of the lowest CFU counts overall. This contradicts the gene copy numbers g^{-1} feces for the corresponding *ermB* and *ermF* genes which are 3rd and 4thhighest after *tet*(O) and *tet*(W) (Figs 2 and 4). A tentative conclusion from this is that the*ermB* and *ermF* genes reside in bacteria found in the intestines other than coliform bacteria as suggested by [31,41–46].

The CFU estimates of the individual animal pool pen samples were found not to represent an average of the non-pooled individual animal samples. This means that, at pen levels, the pooled samples from individual animals were not representative for the individual animals. There were also variations when comparing different sampling methods between pens (Fig 1, bottom) making it difficult to find differences between pens using coliform CFU counts. This is likely due to the large relative standard deviation found for each coliform CFU estimate (Fig 3). Therefore, only qPCR gene copy number g^{-1} feces were used to assess the different sampling and pooling methods.

The qPCR assays revealed interesting differences when assessing whether sampling and pooling strategies within a pig stable were representative of individual animal sampling. Pen 1 consistently had lower gene copy estimates for *ermB*, *ermF*, *tet*(O), and *tet*(W) compared to pens 2–4 with *tet*(C) also having lower estimates in pen 2 (Fig 2 top). These differences in gene copy number g^{-1} feces between pens were found statistically significant, and could mean that the antibiotic resistance genes do not easily spread between pens. The apparent variation in *tet*(A) estimates seen in Fig 2 (top) in pens 1 and 3 is caused by the graph only illustrating results from positive animals. Thus, few animals had high gene copy numbers (above $1x10^6$), where for half of them, one of the technical replicates was below the LOQ and therefore had no effect on the graph. This was solely seen for *tet*(A).

Bibbal et al. [39] monitored the bla_{TEM} excretion in pigs and found that the fecal excretion ofbla_{TEM} genes showed large, individual day-to-day fluctuations [39]. Similar fluctuations in gene excretion could account for variations when quantifying antibiotic resistance genes. Antibiotic resistance is dynamic as its spread and maintenance is subject to fluctuations in host organism migration and/or persistence, antibiotic gene migration, and presence of selection pressure [40]. When collecting rectal samples at a single time point, the level found in all the individual animals within a pen may depend on the time a single animal within the pen has excreted the specific gene. The individual animal pool pen samples are more uniform than the individual animal sampling, as all of the animals and their respective antibiotic resistance levels are represented (Fig 2, top). If a single fecal sample with high levels of antibiotic resistance levels is included in a pool, it will mask the samples containing lower gene levels. On the other hand, if there also are sufficient fecal samples with low gene levels then they will dilute the high level sample [47].

For the majority of the genes, the individual animal pen pool resulted in lower gene copy number estimates when compared to the pen floor samples for each corresponding pen (Fig 2 and S1 Fig). This could be due to the dilution effect of the increased volume in the individual animal pen pools which were composed from >19 samples compared to the pen floor samples that consisted of 5 individual samples [48]. Furthermore, pen floor samples are collected from older feces that has resided on the ground permitting liquid evaporation and run off from the feces. This may result in a higher concentration of resistance genes in the sampled portion of the pen floor fecal sample and can be an advantage when quantifying low prevalence genes. This is illustrated by the fact that *tet*(B) was below the LOQ for the individual animal samples in 50% of the pens, while the pen floor sample from pen 4 and the pen floor pool stable samples were both positive for *tet*(B). In contrast, the individual shoe cover samples were negative for *ermF* while the shoe cover pool stable and individual animal samples from pens 1–4 were positive. Overall, this indicates that the shoe cover pool stable and the pen floor pool stable samples may be used instead of sampling individual animals.

4.3 Pooling strategies at herd level

During sampling 2, 5 stables were sampled from the same herd as sampling 1. An interesting observation was that, for some genes the shoe cover samples were lower than the pen floor samples (*ermB*, *ermF*, *tet*(O), *tet*(W)) but were higher for *sulI*, *sulII*, *tet*(A), and*tet*(M). This was observed both in sampling 1 (Fig 2 bottom) and the stable and herd pools from sampling 2 (Fig 4). The shoe cover samples varied in how much roughage they collected. Thus, if the shoe cover samples with lower gene copy number estimates had more roughage, the sample would weigh more without the entire weight being attributable to feces. Consequently, this could result in low gene copy number g^{-1} feces. In contrast, the shoe cover with high gene copy number estimates suggest that the antibiotic resistance genes represented in the sample may depend on which sampling method is used, as different bacteria harbor antibiotic resistance genes while residing in a specific fecal fraction. The gastrointestinal tract is a complex ecosystem containing at least 400 different bacterial species residing in regional habitats [49]. Hence, the shoe cover samples may be capable of collecting fractions of feces that pen floor or individual fecal samples cannot as the shoe cover samples were both saturated with liquid and covered with feces after sample collection. Furthermore, the entire pen floor is covered during shoe cover sampling thus increasing the likelihood of collecting a fecal sample positive for a given gene.

Several genes tested positive in at least one non-pooled sample type which then turned negative after pooling (sampling 1 *sulI*, *sulII*, *tet*(A); sampling 2 *tet*(A), *tet*(B), *tet*(C), *sulII*). Pooling may increase the risk of a sample becoming negative, especially if the gene copy number g^{-1} feces initially are relatively low. If there are sufficient numbers of negative samples included in the pool then the low prevalence gene concentration is diluted potentially resulting in levels under the assay's quantification limit [48]. The sensitivity of a specific assay is therefore dependent on the gene prevalence, the number of samples included in the pool, the gene concentration in samples collected from positive animals, and the quantification limit of the assay [47,48]. A balance in the mentioned parameters could explain the apparent consistency in gene copy number

g^{-1} feces between slurry tank samples, pen floor and shoe cover herd pools for the majority of the assays.

When choosing a sampling method for antibiotic resistance determination at herd level by qPCR, we recommend either pen floor samples or shoe cover sampling. Both sampling types were able to quantify antibiotic resistance genes in swine feces. Pen floor samples were easily attainable and are representative when pooled at the stable level, but several were negative when pooled at herd level (Fig 4 bottom). In contrast, the shoe cover samples include the entire pen, thus representing more animals and increasing the likelihood of finding antibiotic resistant determinants when present. Furthermore, fecal fractions not represented in pen floor samples and individual animal samples may be represented in the shoe cover samples. However, the shoe cover samples should not be pooled as the herd pools were negative for several of the assays. The slurry tank samples were also promising; their gene copy levels were consistent with those quantified in the pen floor and shoe cover samples. The slurry tank contains feces from the entire herd from a time period of approximately 6 months and may therefore give a better estimate of the herd antibiotic resistance levels. Further studies should be conducted where a series of slurry tank samples are collected from several pig herds over a longer time period, for example 6 months, in order to clarify the dynamics of antibiotic resistance genes in slurry tanks.

This study has utilized two different approaches to antibiotic resistance surveillance to assess different sampling and pooling methods of swine fecal sample collection that yielded two different endpoints, namely 14 qPCR assays to quantify antibiotic resistance genes in swine feces and CFU counts of coliform bacteria in the same samples. To our knowledge, this is the first study that tests sampling and pooling strategies for antibiotic resistance surveillance using qPCR determination of antibiotic resistance in total DNA extracted from swine feces [12,50,51]. It is necessary to understand the quantified antibiotic resistance gene levels in order to define if certain levels pose a potential risk or if they represent the specific farm. In order to do so, studies including parallel analysis of antibiotic resistance using several methods should be conducted across several pig farms.

Our results indicate that there is a great deal of variation in the antibiotic gene abundance within individual animals, pens, stables, and herds regardless of the sampling method. This variation could be systematically evaluated in greater detail using pen floor and/or shoe covering sampling methods supplemented with parallel slurry tank sampling.

ACKNOWLEDGMENTS

We thank Yvonne Agersø, Annette M. Hammerum, Stefan Schwarz, and Luca Guardabassi for providing strains used in this study, Katja Kristensen, Tony Boennelycke, and Peter Damburg for providing technical assistance, The Public Health Agency of Sweden for providing the 16S rDNA qPCR assays, Inge Larsen for establishing contact to the swine herd in this study, and finally Henrik Nielsen for permitting us to collect samples from his pig herd.

AUTHOR CONTRIBUTIONS

Conceived and designed the experiments: GVS AM JEO ØA MS LEC. Performed the experiments: GVS JEO ØA AM. Analyzed the data: LEC GVS ØA. Contributed reagents/materials/analysis tools: GVS JEO ØA. Wrote the paper: GVS AM JEO ØA MS LEC.

REFERENCES

1. Aarestrup FM. Association between the consumption of antimicrobial agents in animal husbandry and the occurrence of resistant bacteria among food animals. Int J Antimicrob Ag 1999;12(4):279–285.

2. McGowan JE Jr. Economic Impact of Antimicrobial Resistance. Emerg Infect Dis 2001;7(2):286. pmid:11294725

3. World Health Organization. Surveillance standards for antimicrobial resistance. 2002;WHO/CDS/CSR/DRS/2001.5.

4. van den Bogaard AE, Stobberingh EE. Epidemiology of resistance to antibiotics links between animals and humans. Int J Antimicrob Agents 2000;14(4):327–335. pmid:10794955

5. Sunde M, Fossum K, Solberg A, Sørum H. Antibiotic resistance in *Escherichia coli* of the normal intestinal flora of swine. Microb Drug Resist 1998;4(4):289–299. pmid:9988047

6. Schwarz S, Kehrenberg C, Walsh TR. Use of antimicrobial agents in veterinary medicine and food animal production. Int J Antimicrob Agents 2001; 6;17(6):431–437. pmid:11397611

7. Alali WQ, Scott HM, Norby B, Gebreyes W, Loneragan GH. Quantification of the*bla*$_{CMY-2}$ in feces from beef feedlot cattle administered three different doses of ceftiofur in a longitudinal controlled field trial. Foodborne Pathog Dis 2009;6(8):917–924. doi: 10.1089/fpd.2009.0271. pmid:19622032

8. Fluit AC, Visser MR, Schmitz FJ. Molecular Detection of Antimicrobial Resistance. Clin Microbiol Rev 2001;14(4):836–871. pmid:11585788 doi: 10.1128/cmr.14.4.836-871.2001

9. Yu Z, Michel FC Jr, Hansen G, Wittum T, Morrison M. Development and application of real-time PCR assays for quantification of genes encoding tetracycline resistance. Appl Environ Microbiol 2005;71(11):6926–6933. pmid:16269727

10. Chen J, Yu Z, Michel FC Jr, Wittum T, Morrison M. Development and application of real-time PCR assays for quantification of erm genes conferring resistance to macrolides-lincosamides-streptogramin B in livestock manure and manure management systems. Appl Environ Microbiol 2007;73(14):4407–4416. pmid:17496134 doi: 10.1128/aem.02799-06

11. Knapp CW, Zhang W, Sturm BSM, Graham DW. Differential fate of erythromycin and beta-lactam resistance genes from swine lagoon waste under different aquatic conditions. Environ Poll 2010; 158(5):1506–1512.

12. Skov MN, Carstensen B, Tornøe N, Madsen M. Evaluation of sampling methods for the detection of *Salmonella* in broiler flocks. J App Microbio 1999;86:695–700.

13. Wegener HC, Hald T, Lo Fo Wong D, Madsen M, Korsgaard H, Bager F, et al.*Salmonella* Control Programs in Denmark. Emerg Infect Diseases 2003;(97):774–780.

14. Fødevareministeriet. Bekendtgørelse om salmonella hos svin. 2012;BEK 404 af 08/05/2012.

15. Caprioli A, Busani L, Martel JL, Helmuth R. Monitoring of antibiotic resistance in bacteria of animal origin: epidemiological and microbiological methodologies. Int J Antimicrob Agents 2000; 14(4):295–301. pmid:10794950

16. DANMAP. DANMAP 2012—Use of antimicrobial agents and occurrence of antimicrobial resistance in bacteria from food animals, food and humans in Denmark. 2013(ISSN 1600-2032).

17. Chen C, Nace GW, Irwin PL. A 6×6 drop plate method for simultaneous colony counting and MPN enumeration of *Campylobacter jejuni*, *Listeria monocytogenes*, and *Escherichia coli.* J Microbiol Methods 2003;11;55(2):475–479. pmid:14529971

18. Hoelzel CS, Harms KS, Kcochenhoff H, Kunz A, Mcoller C, Meyer K, et al. Phenotypic and genotypic bacterial antimicrobial resistance in liquid pig manure is variously associated with contents of tetracyclines and sulfonamides. J Appl Microbiol 2010;108(5):1642–1656.

19. Kazimierczak KA, Scott KP, Kelly D, Aminov RI. Tetracycline Resistome of the Organic Pig Gut. App Environ Microbiol 2009;75(6):1717–1722.

20. CheeSanford JC, Mackie RI, Koike S, Krapac IG, Lin YF, Yannarell AC, et al. Fate and transport of antibiotic residues and antibiotic resistance genes following land application of manure waste. J Environ Qual 2009;38(3):1086–1108.

21. De Leener E, Martel A, Decostere A, Haesebrouck F. Distribution of the erm (B) Gene, tet racycline Resistance Genes, and Tn 1545-like Transposons in Macrolide-and Lincosamide-Resistant Enterococci from Pigs and Humans. Microbial Drug Resistance 2004;10(4):341–345. pmid:15650380

22. Garofalo C, Vignaroli C, Zandri G, Aquilanti L, Bordoni D, Osimani A, et al. Direct detection of antibiotic resistance genes in specimens

of chicken and pork meat. Int J Food Microbiol 2007;113(1):75–83. pmid:17005283

23. Hammerum AM, Sandvang D, Andersen SR, Seyfarth AM, Porsbo LJ, FrimodtMoller N, et al. Detection of sul1, sul2 and sul3 in sulphonamide resistant *Escherichia coli*isolates obtained from healthy humans, pork and pigs in Denmark. Int J Food Microbiol 2006; 1;106(2):235–237. pmid:16216373

24. Byrne-Bailey KG, Gaze WH, Kay P, Boxall AB, Hawkey PM, Wellington EM. Prevalence of sulfonamide resistance genes in bacterial isolates from manured agricultural soils and pig slurry in the United Kingdom. Antimicrob Agents Chemother 2009; 53(2):696–702.

25. Moodley A, Guardabassi L. Transmission of IncN plasmids carrying *bla*$_{CTX\text{-}M\text{-}1}$between commensal *Escherichia coli* in pigs and farm workers. Antimicrob Agents Chemother 2009;53(4):1709–1711.

26. Agerso Y, Aarestrup FM, Pedersen K, Seyfarth AM, Struve T, Hasman H. Prevalence of extended-spectrum cephalosporinase (ESC)-producing *Escherichia coli* in Danish slaughter pigs and retail meat identified by selective enrichment and association with cephalosporin usage. J Antimicrob Chemother 2012;67(3):582–588.

27. TJ, Plewniak F, Jeanmougin F, Higgins DG. The CLUSTAL_X windows interface: flexible strategies for multiple sequence alignment aided by quality analysis tools. Nucleic Acids Res 1997;15;25(24):4876–4882. pmid:9396791

28. Aminov RI, Chee-Sanford JC, Garrigues N, Teferedegne B, Krapac IJ, White BA, et al. Development, validation, and application of PCR primers for detection of tetracycline efflux genes of gram-negative bacteria. Appl Environ Microbiol 2002;68(4):1786–1793. pmid:11916697

29. Boeckelmann U, Doerries HH, Ayuso-Gabella MN, Salgot de Marcay M, Tandoi V, Levantesi C, et al. Quantitative PCR Monitoring of Antibiotic Resistance Genes and Bacterial Pathogens in Three Eurpean Artificial Groundwater Recharge Systems. Appl Environ Microbiol 2009;75(1):154–163.

30. Smith MS, Yang RK, Knapp CW, Niu YF, Peak N, Hanfelt MM, et al. Quantification of tetracycline resistance genes in feedlot lagoons

by real-time PCR. Appl Environ Microbiol 2004;70(12):7372–7377. pmid:15574938

31. Jacoby GA. ß-Lactamase Classification and Amino Acid Sequences for TEM, SHV and OXA Extended-Spectrum and Inhibitor Resistant Enzymes. Available at:http://www.lahey.org/Studies/. Accessed 15–08, 2013.

32. Birkett CI, Ludlam HA, Woodford N, Brown DF, Brown NM, Roberts MT, et al. Real-time TaqMan PCR for rapid detection and typing of genes encoding CTX-M extended-spectrum beta-lactamases. J Med Microbiol 2007;56:52–55. pmid:17172517

33. L, Collard JM, Olsen JE, Dalsgaard A. Distribution and in-vitro transfer of tetracycline resistance determinants in clinical and aquatic*Acinetobacter* strains. J Med Microbiol 2000;49:929–936. pmid:11023190

34. Hasman H, Mevius D, Veldman K, Olesen I, Aarestrup FM. beta-Lactamases among extended-spectrum beta-lactamase (ESBL)-resistant *Salmonella* from poultry, poultry products and human patients in The Netherlands. J Antimicrob Chemother 2005; 56(1):115–121. pmid:15941775

35. Archambault M, Petrov P, Hendriksen RS, Asseva G, Bangtrakulnonth A, Hasman H, et al. Molecular characterization and occurrence of extended-spectrum beta-lactamase resistance genes among *Salmonella enterica* serovar Corvallis from Thailand, Bulgaria, and Denmark. Microb Drug Resist 2006;12(3):192–198. pmid:17002546

36. Angen O, Ahrens P, Tegtmeier C. Development of a PCR test for identification of*Haemophilus somnus* in pure and mixed cultures. Vet Microbiol 1998;28;63(1):39–48. pmid:9810620

37. Agerholm JS, Christoffersen AB, Agger JF. Dynamics of relationship between the presence of *Coxiella burnetii* DNA, antibodies, and intrinsic variables in cow milk and bulk tank milk from Danish dairy cattle. J Dairy Sci 2011;12;94(12):5750–5759.

38. Harris S, Jones DB. Optimisation of the polymerase chain reaction. British Journal of Biomedical Science 1997;54:166–173. pmid: 9499593

39. Bibbal D, Dupouy V, Ferre JP, Toutain PL, Fayet O, Prere MF, et al. Impact of three ampicillin dosage regimens on selection of ampicillin resistance in*Enterobacteriaceae* and excretion of *bla*$_{TEM}$ genes in swine feces. Appl Environ Microbiol 2007;73(15):4785–4790. pmid:17557857

40. Koike S, Krapac IG, Oliver HD, Yannarell AC, CheeSanford JC, Aminov RI, et al. Monitoring and source tracking of tetracycline resistance genes in lagoons and groundwater adjacent to swine production facilities over a 3-year period. Appl Environ Microbiol 2007;73(15):4813–4823. pmid:17545324

41. Gniadkowski M, Schneider I, Palucha A, Jungwirth R, Mikiewicz B, Bauernfeind A. Cefotaxime-resistant *Enterobacteriaceae* isolates from a hospital in Warsaw, Poland: identification of a new CTX-M-3 cefotaxime-hydrolyzing beta-lactamase that is closely related to the CTX-M-1/MEN-1 enzyme. Antimicrob Agents Chemother 1998;42(4):827–832. pmid:9559791

42. Heritage J, M'Zali FH, Gascoyne-Binzi D, Hawkey PM. Evolution and spread of SHV extended-spectrum beta-lactamases in Gram-negative bacteria. J Antimicrob Chemother 1999;44(3):309–318. pmid:10511397

43. Chanawong A, M'Zali FH, Heritage J, Lulitanond A, Hawkey PM. SHV-12, SHV-5, SHV-2a and VEB-1 extended-spectrum beta-lactamases in Gram-negative bacteria isolated in a university hospital in Thailand. J Antimicrob Chemother 2001;48(6):839–852. pmid: 11733468

44. Navarro F, Perez-Trallero E, Marimon JM, Aliaga R, Gomariz M, Mirelis B. CMY-2-producing *Salmonella enterica*, *Klebsiella pneumoniae*, *Klebsiella oxytoca*, *Proteus mirabilis* and *Escherichia coli* strains isolated in Spain (October 1999-December 2000). J Antimicrob Chemother 2001;48(3):383–389. pmid:11533003

45. Baraniak A, Fiett J, Sulikowska A, Hryniewicz W, Gniadkowski M. Countrywide spread of CTX-M-3 extended-spectrum beta-lactamase-producing microorganisms of the family *Enterobacteriaceae* in Poland. Antimicrob Agents Chemother 2002; 46(1):151–159. pmid:11751126

46. Li XZ, Mehrotra M, Ghimire S, Adewoye L. beta-Lactam resistance and beta-lactamases in bacteria of animal origin. Vet Microbiol 2007;15;121(3–4):197–214. pmid:17306475

47. Muñoz-Zanzi CA, Johnson WO, Thurmond MC, Hietala SK. Pooled-sample testing as a herd-screening tool for detection of bovine viral diarrhea virus persistently infected cattle. J Vet Diagn Invest 2000;12(3):195–203. pmid:10826831

48. Muñoz-Zanzi C, Thurmond M, Hietala S, Johnson W. Factors affecting sensitivity and specificity of pooled-sample testing for diagnosis of low prevalence infections. Prev Vet Med 2006; 74(4):309–322. pmid:16427711

49. Falk PG, Hooper LV, Midtvedt T, Gordon JI. Creating and Maintaining the Gastrointestinal Ecosystem: What We Know and Need to Know from Gnotobiology. Microbio Molec Bio Rev 1998;62(4):1157–1170.

50. Dunlop RH, McEwen SA, Meek AH, Friendship RM, Black WD, Clarke RC. Sampling considerations for herd-level measurement of faecal *Escherichia coli* antimicrobial resistance in finisher pigs. Epi and Infect 1999;122(3):485–496.

51. Broens EM, Graat EAM, Engel B, van Oosterom RAA, van de Giessen AW, van der Wolf PJ. Comparison of sampling methods used for MRSA-classification of herds with breeding pigs. Vet Microbiol 2011;147(3–4):440–444.

CHAPTER 8

Longitudinal study on the colonisation and transmission of methicillin-resistant *Staphylococcus aureus* in pig farms

Patrick Daniel Bangerter a , Xaver Sidler b , Vincent Perreten a , Gudrun Overesch a, *

a Institute of Veterinary Bacteriology, Vetsuisse Faculty, University of Bern, Länggassstrasse 122, CH-3012 Bern, Switzerland

b Department of Farm Animals, Division of Swine Medicine, Vetsuisse Faculty, University of Zurich, Winterthurerstrasse 260, CH-8057 Zurich, Switzerland

ABSTRACT

Knowledge about the dynamics of methicillin-resistant *Staphylococcus aureus* (MRSA) in pigs lacks detail at the level of individual animal. The aim of our study was therefore to determine the colonisation status of MRSA in individual pigs from birth to slaughter in order to gain a better understanding of substantial factors involved in transmission. Two farrow-to-finish and two grow-to-finish herds were included in the study. A total of 1728 nasal swabs from 390 pigs and 592 environmental wipes were collected at 11 different time points.

Intermittent colonisation throughout the entire production cycle was conspicuous in the tracking of MRSA in individual pigs. Almost all pigs from a MRSA-positive herd changed MRSA status several times, which implies that pigs are transiently rather than permanently colonised. We highly recommend the definition of MRSA status at herd level rather that at the level of the individual pig when considering prevention measures against MRSA. Therefore, to avoid the further spread of MRSA in countries with moderate prevalence, such as in Switzerland, defining farms as MRSA positive or MRSA negative and allowing the trade of pigs only within herds of the same status seems feasible. This will also be important for combating the further dissemination of livestock-associated (LA)-MRSA into healthcare facilities and the community via humans who have close contact with animals.

KEYWORDS

LA-MRSA; Transportation; Slaughterhouse; Public health; Prevention; Swine

1. INTRODUCTION

The rapid spread of LA-MRSA in pigs and farm animals worldwide has raised major public health concerns (Crombé et al., 2013, Verkade and Kluytmans, 2014 and Voss et al., 2005). Colonised animals may act as a MRSA reservoir not only for livestock but also for humans with close contact to animals, i.e., farmers and veterinarians. As a consequence, higher colonisation rates and cases of infections have been reported in these professions at risk (Lewis et al., 2008, van Rijen et al., 2008 and Wettstein Rosenkranz et al., 2014).

In 2009, official monitoring was launched for MRSA in pigs at slaughterhouses in Switzerland. The prevalence of MRSA in 2009 was very low at 2% (95% CI 0.9–3.9) but reached 20.8% (95% CI 16.7–25.45) in 2013 (Büttner et al., 2014 and Overesch et al., 2012).

To date, little is known about the dynamics of MRSA in pigs because only a few longitudinal studies have been conducted. Those longitudinal studies that have been conducted examined the MRSA status in pigs mostly until slaughter age (Broens et al., 2012, Burns et al., 2014 and Verhegghe et al., 2013) or even just until the age of 70 days (Weese et al., 2011). Moreover, these studies did not provide results from individual pigs. Other studies examined only one MRSA-positive farm (Burns et al., 2014 and Weese et al., 2011), and the results are unlikely to be generally applicable. Other researchers (Broens et al., 2011) considered the prevalence of MRSA before and after transportation of the pigs from farm to abattoir, but they did not examine the changes in the MRSA status of individual pigs at the farm while simply analysing pooled samples, from which individual changes could not be demonstrated. The dramatic increase of MRSA in Swiss slaughter pigs during recent years necessitates the introduction of measures to combat the further spread of MRSA in the Swiss pig population. However, until now, there have been no precise studies of the individual colonisation dynamics of MRSA throughout each pig production stage; these studies are needed to gain a better understanding of the substantial factors considering the prevention of the spread of MRSA and to identify targets for possible intervention measures. For that reason, we selected MRSA-positive and negative farms with different management practices, such as all in/all out and continuous animal flow systems and determined the MRSA status in the individual pig from birth throughout each production stage, up to and including transport and slaughtering.

2. MATERIALS UND METHODS

2.1. Farm characteristics and animals

Pigs from four Swiss pig farms were recurrently tested for the presence of MRSA between May and December 2014 during a production cycle of approximately 150–175 days, as well as additional pigs from three other farms on transportation to slaughterhouses. We chose two farrow-to-finish farms (ff-I, ff-II) as well as two grow-to-finish farms (gf-I, gf-II) for analysis, to determine if any variances between the different management systems exists. One farrow-to-finish farm was chosen as a MRSA-negative control farm whereas the other farrow-to-finish farm was MRSA-positive. Furthermore, one of the grow-to-finish farms was purchaser of grower pigs from a farrow-to-finish study-farm while the other grow-to-finish farm was not associated to any of the farms.

Farm ff-I consisted of 75 sows, 50 replacement gilts, two farrowing rooms for 18 sows in each room, one weaner accommodation and a finishing unit with a capacity for 200 fattening pigs. The farm had a three-week batch monitoring system. One cohort of ten pregnant sows approaching delivery was selected for starting sampling and placed after washing in a cleaned and disinfected farrowing room. Four sows were placed in pens with possible direct contact to the neighbouring pen through an open fence. The other six sows had no contact.

Farm gf-I purchased grower pigs from farm ff-I but also from other breeders. The farm had seven finishing pens with a capacity for 280 fattening pigs. One finishing pen containing 37 fattening pigs grown on farm ff-I was selected for sampling.

Farm ff-II consisted of 42 sows, one farrowing room with 12 farrowing pens, one weaner accommodation and a finishing unit with a capacity for 250 fattening pigs. Replacement gilts were purchased. The farm had no regular batch-monitoring system and routinely used post-weaning prophylactic feed supplementation with lincomycin and spectinomycin for three weeks. One cohort of three pregnant sows approaching delivery was selected for starting sampling. All three sows were placed in pens with possible direct contact to the neighbouring pen through an open fence. Five other pens were also occupied. The status of those sows was unknown.

Farm gf-II had one finishing pen with a capacity for approximately 90 fattening pigs. A total of 87 fattening pigs purchased from one breeder were selected for sampling.

More details of the farms are given in Table 1. Samples and time points are listed in Table 2.

2.2. Collection of samples

Nasal swabs from individual pigs were taken at the different time points during a production cycle up to and including slaughtering. When indicated, additional environmental wipes were also taken (Table 2). At the two farrow-to-finish farms (ff-I, ff-II), sows were sampled three times and their offspring nine times. At approximately four to five weeks of age, the piglets were merged into new groups. Each group was housed in a separate pen in one room in the weaner accommodation. At the age of nine to ten weeks at tp7, the grower pigs were merged into new groups and moved to first stage finishing. At tp8, at approximately 14 weeks of age, fattening pigs were moved to second stage finishing.

Table 1. Characteristics of the farms.

Farm no	MRSA status[a]	Operation type	Other livestock on farm	Antimicrobial group treatment[b]
ff-I	Positive	Farrow- to-finish farm Continuous flow	No	No
gf-I	Positive	Grow-to-finish farm Continuous flow	Yes (calves)	No
ff-II	Negative	Farrow-to-finish farm Continuous flow	No	Yes
gf-II	Positive	Grow-to-finish farm All in–all out flow	Yes (dairy cows)	No

[a]According to previous screening.
[b] During the production cycle.

Table 2. Time points of sampling, type of sample, age group and age of the pigs, sampled compartment and farm on which samples were taken.

Time point	Type of sample	Age group	Age	Sampled compartment	Farm
1	Nasal	Sows	1 week prior to farrowing		ff-I, ff-II
	Environment			Farrowing[a]	ff-I, ff-II
2	Nasal	Piglets/Sows	Within 24 h post-natal		ff-I, ff-II
3	Nasal	Piglets	5–14 days		ff-I, ff-II
4	Nasal	Weaned piglets/Sows	24–38 days		ff-I, ff-II
	Environment			Farrowing[b]/Weaning[a]	ff-I, ff-II
5	Environment[c]			Farrowing[a]	ff-I
6	Nasal	Weaned piglets	5–7 weeks		ff-I, ff-II
7	Nasal	Growers	8–11 weeks		ff-I, ff-II, gf-I, gf-II
	Environment			Weaning[b]	ff-I, ff-II
	Environment			Fattening[a]	ff-I, ff-II, gf-I, gf-II
8	Nasal	Finishers	15–19 weeks		ff-I, ff-II, gf-I, gf-II
	Environment			Fattening[b]	ff-I, ff-II, gf-I, gf-II
	Environment			Finishing[a]	ff-I, ff-II
9 i	Nasal	Finishers	21–25 weeks		ff-I, ff-II, gf-I, gf-II
	Environment			Fattening[b]	gf-I, gf-II
	Environment			Finishing[b]	ff-I, ff-II
	Environment			Lorry[a]/Abattoir[a]	ff-I, ff-II, gf-I, gf-II
9 ii	Nasal	Finishers	21–25 weeks		ff-I, gf-I, gf-II
	Environment			Lorry[b]	ff-I, ff-II, gf-I, gf-II
9 iii	Nasal	Finishers	21–25 weeks		ff-I, ff-II, gf-I, gf-II
	Environment			Abattoir[b]	ff-I, ff-II, gf-I, gf-II

[a] Before use.
[b] After use.
[c] Environmental wipes from the same farrowing pens, two days after weaning afterwards cleaning and disinfection.

On the two grow-to-finish farms (gf-I, gf-II) grower pigs were purchased and housed in the finishing pens, where sampling began. They were sampled five times.

At the end of fattening, on all four farms (ff-I, ff-II, gf-I, gf-II), samples were taken at three different times, i.e., before (tp9i) and after (tp9ii) transportation and after stunning or bleeding (tp9iii). Slaughter pigs from the four farms were transported to three different commercial abattoirs (slaughterhouse sh-I, sh-II and sh-III), namely farm ff-I and gf-I to slaughterhouse sh-I, farm ff-II to slaughterhouse sh-II and farm gf-II to slaughterhouse sh-III. The lairages at slaughterhouse sh-I and sh-III were unused and clean, whereas the lairages at slaughterhouse sh-II had already been used earlier that day and were therefore not clean.

On transportation to the abattoir from farm ff-1, additional pigs (n = 42) from one other farm were picked up by the same lorry. Initial nasal samples were taken from these pigs on the farm immediately before transportation (tp9i). During transport from farm gf-II, additional pigs from two other farms (farm 1, n = 56; farm 2, n = 21) also were picked up by the same lorry. Initial nasal samples from these pigs were taken on one day and two days before transport (tp9i). Additional pigs from the other farms were located in separate lorry sections, but contact between the pigs was possible. Moreover, for the transportation of batch ff-I and batch gf-I, 10 and 5 supplemental pigs, respectively, which were not part of the study, were transported and tested as well but only at tp9i, 9ii and 9iii.

Nasal samples were collected using transport swabs (Transwab® Amies MW172, MWE Medical Wire, Corsham, England and Uni-Ter Amies CLR, Meus S.r.l., Piove Di Sacco, Italy) from both nares of the pigs. Environmental wipes (lairs, wall, watering place, manger and steel parts) were collected from the farrowing pens, the weaner accommodation, the fattening units and the lairages at the slaughterhouse, as well as from the lorries, using wipes (Triko-Tex® 18 × 32 cm, Chicopee Europe, Katwijk, The Netherlands) moistened with distilled water. Samples were taken by wiping the surface of the steel parts, each part with one wipe, and the wall at the height of the snout. In each box, one wall was selected for sampling. Each wipe was individually placed in a sterile Stomacher bag. Depending on the transfer or death of the pigs, the number of sampled pigs varied over time. 144 pigs were not tested throughout the complete period of the study for different reasons like death, slaughtered later, used for breeding. Details for individual exclusion are given in Fig. 1, Fig. 2, Fig. 3, Fig. 4, Fig. 5 and Fig. 6.

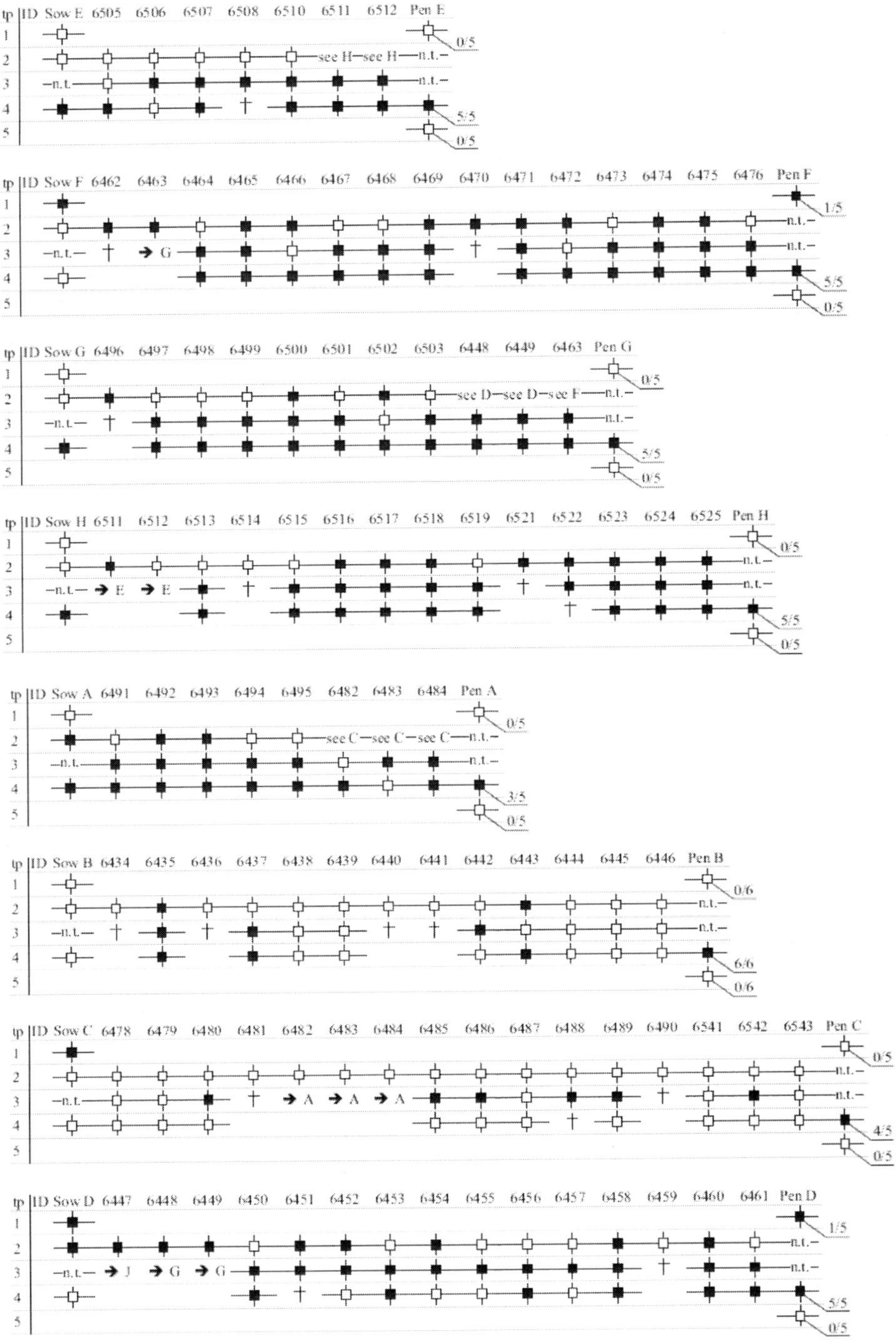
tp ID Sow E 6505 6506 6507 6508 6510 6511 6512 Pen E
see H—see H—n.t.
0/5 5/5 0/5
tp ID Sow F 6462 6463 6464 6465 6466 6467 6468 6469 6470 6471 6472 6473 6474 6475 6476 Pen F
→ G
1/5 5/5 0/5
tp ID Sow G 6496 6497 6498 6499 6500 6501 6502 6503 6448 6449 6463 Pen G
see D—see D—see F—n.t.
0/5 5/5 0/5
tp ID Sow H 6511 6512 6513 6514 6515 6516 6517 6518 6519 6521 6522 6523 6524 6525 Pen H
→ E → E
0/5 5/5 0/5
tp ID Sow A 6491 6492 6493 6494 6495 6482 6483 6484 Pen A
see C—see C—see C—n.t.
0/5 3/5 0/5
tp ID Sow B 6434 6435 6436 6437 6438 6439 6440 6441 6442 6443 6444 6445 6446 Pen B
0/6 6/6 0/6
tp ID Sow C 6478 6479 6480 6481 6482 6483 6484 6485 6486 6487 6488 6489 6490 6541 6542 6543 Pen C
→ A → A → A
0/5 4/5 0/5
tp ID Sow D 6447 6448 6449 6450 6451 6452 6453 6454 6455 6456 6457 6458 6459 6460 6461 Pen D
→ J → G → G
1/5 5/5 0/5
n.t.

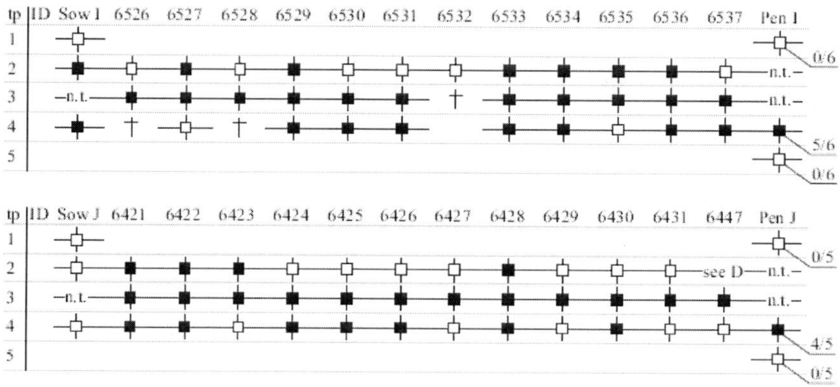

Figure 1. MRSA status of sows and offspring on farm ff-I. □, MRSA-negative; ■, MRSA-positive; †, deceased; →, piglet moved; ID, ear tag; tp, time point; n.t., not tested; 2/15, environmental samples (*n* positive/*N* total).

2.3. Sample processing for MRSA isolation

Samples were transported within three hours at ambient temperature to the laboratory, with the exception of nasal swabs from farm ff-1 at tp2. Samples from new-born piglets were collected at 5 °C for a maximum of six days and afterwards transported under cooled conditions (<12 °C) to the laboratory. After delivery, swabs were transferred into tubes containing 10 ml Mueller Hinton Broth supplemented with 6.5% NaCl. A total of 50 ml Mueller Hinton Broth supplemented with 6.5% NaCl was added to each Stomacher bag containing environmental wipes and homogenised in a Stomacher® 400 circulator (Seward Ltd., UK) for 15 s at 230 RPM. The samples were incubated aerobically at 37 °C for 24 h while being shaken. One ml from each pre-enrichment was inoculated into 9 ml tryptone soy broth containing 3.5 mg/l cefoxitin and 75 mg/l aztreonam, and further incubated aerobically at 37 °C for 24 h. Ten microliters was then spread onto MRSA selective agar plates (BBL™ CHROMagar™ MRSA; Becton Dickinson, Franklin Lakes, NJ, USA), which were incubated at 37 °C for 24 h. Pink to mauve-coloured colonies were regarded as suspicious, and one presumptive colony from each plate was cultivated onto tryptone soy agar plates containing 5% sheep blood (TSA-SB) (Oxoid Ltd, Basingstoke, England) at 37 °C for 24 h. *S. aureus* was identified by matrix-assisted laser desorption/ionisation time-of-flight mass spectroscopy (MALDI TOF MS) (Biotyper 3.0, Bruker) using the direct transfer protocol recommended by the manufacturer. The identification of all MRSA isolates was confirmed by PCR targeting the *mecA* gene, which encodes for methicillin-resistance (Stegger et al., 2011). Positive (*S. aureus* MRSA K/M1474/08 laboratory collection) and negative (*S. aureus* ATCC 25923) control strains were included. The obtained MRSA isolates were subsequently stored at −80 °C in trypticase soy broth supplemented with 30% glycerol for further examination.

2.4. Statistical methods

Prevalences in the text were calculated on the number of MRSA positive pigs based on the total number of pigs at the particular time point. Rates of changes in the text and Table 3 were calculated as number of changes from MRSA negative to positive based on pigs that were negative at the first time point and from MRSA positive to negative based on MRSA positive pigs at the respective time point. At last, rates of pigs without change of the MRSA status based on the total number of pigs at the first time point were calculated. Values are given in percent. All statistics were performed using the NCSS 10 statistical software (2015, NCSS, LLC. Kaysville, Utah, USA, ncss.com/software/ncss).

3. RESULTS

In total, 1728 nasal swabs from 390 pigs and 592 environmental wipes were collected from the four studied farms, as well as from three additional farms from which samples were taken only in pigs being transported. Individual results of the animal and environmental samples are summarised in Fig. 1, Fig. 2, Fig. 3, Fig. 4, Fig. 5 and Fig. 6. Moreover, Table 3 shows the percentage changes of the MRSA status of the pigs for every transition from one time point to another for every farm.

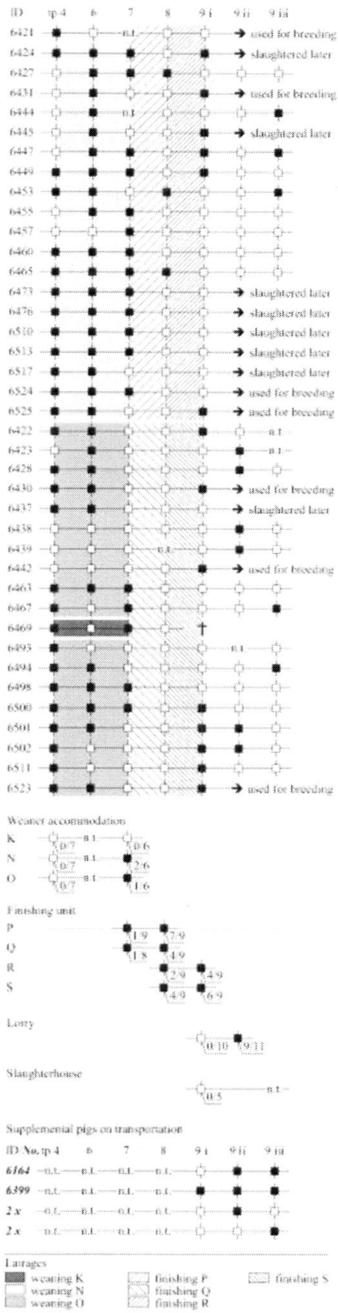

Figure 2. Fattening pigs weaned and fattened on farm ff-I. ***6164-6399***, supplemental fattening pigs from farm ff-I which were not part of the study, sampled at tp 9 only. □, MRSA-negative; ■, MRSA-positive; †, deceased; ID, ear tag; tp, time point; n.t., not tested; 2/15, environmental samples (*n* positive/*N* total).

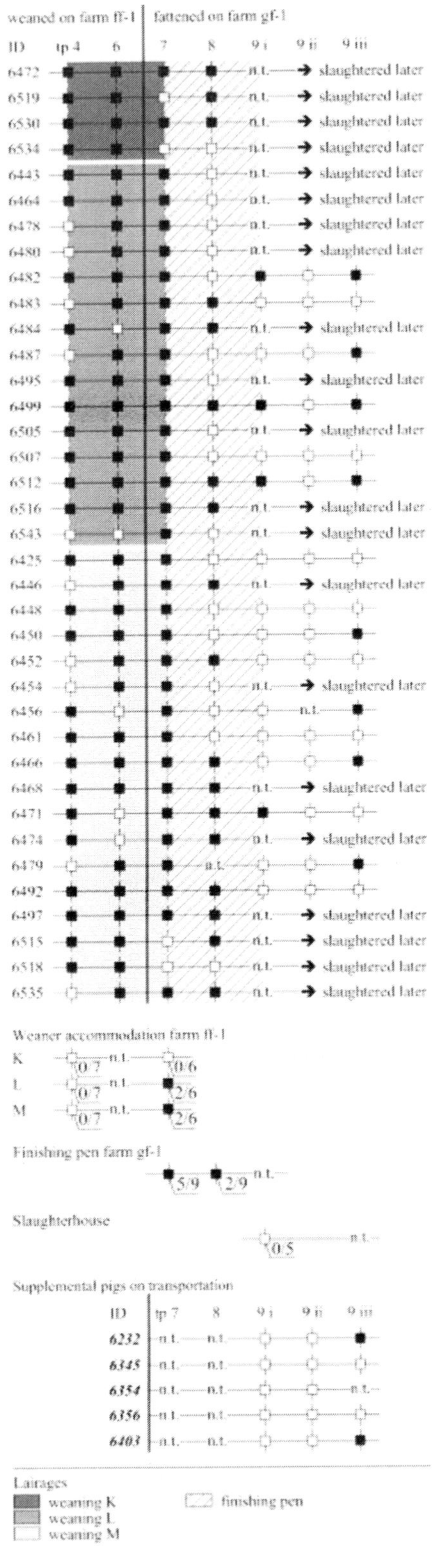
weaned on farm ff-1
fattened on farm gf-1
ID
tp 4
6
7
8
9 i
9 ii
9 iii
slaughtered later
n.t.
Weaner accommodation farm ff-1
K
L
M
0/7
0/6
2/6
Finishing pen farm gf-1
5/9
2/9
Slaughterhouse
0/5
Supplemental pigs on transportation
6232
6345
6354
6356
6403
Lairages
weaning K
weaning L
weaning M
finishing pen

Figure 3. Fattening pigs weaned on farm ff-I and purchased for fattening to farm gf-I. ***6232-6403***, supplemental fattening pigs from farm gf-I which were purchased from the same breeder (ff-I) but not part of the study. Sampled at tp 9 only. □, MRSA-negative; ■, MRSA-positive; ID, ear tag; tp, time point; n.t., not tested; 2/15, environmental samples (*n* positive/*N* total).

On **farm ff-I**, of ten sows (A–J) sampled before farrowing at tp1, three were MRSA positive (sows C, D and F) and seven were MRSA negative (sows A, B, E, G, H, I and J) (Fig. 1). Of the three MRSA-positive sows, two (sows C and F) had showed changed MRSA status 24 h after birth (tp2) to negative and stayed negative until the piglets were weaned at approximately 28 days (tp4). Only two sows (sows B and J) were screened as negative at all three sampling moments. After up to ten days of life (tp3), the majority of the piglets tested MRSA positive, i.e., 84.2% (85/101), as a consequence of high percentage changes of the MRSA status from tp2 to tp3 (80.3%) (Table 3).

After weaning, the environmental wipes from the pens were almost all MRSA positive, whereas wipes, after the farrowing pens had been cleaned, were all negative.

At approximately 28 days of age (tp4), pigs (n = 95) were merged in weaner groups K, L, M, N and O and penned in the weaner accommodation (Fig. 2 and Fig. 3). Thirty-one of the weaned pigs that were sampled 3 times at the weaner accommodation (tp4, 6 and 7) remained MRSA positive at all tp sampled (41.9%; 31/74), but all other pigs (54.1%; 40/74) changed their MRSA status once or even twice. The prevalence at tp4 (before rehousing), 6 and 7 was 69.5% (66/95), 80.3% (61/76) and 68.9% (51/74), respectively. Environmental wipes from the weaner accommodation were all MRSA negative previous to penning, whereas after holding pigs, nearly all pens tested MRSA positive (Fig. 2 and Fig. 3).

Figure 4. Fattening pigs on farm ff-II. □, MRSA-negative; ■, MRSA-positive; †, deceased. ID, ear tag; tp, time point; n.t., not tested; 2/15, environmental samples (*n* positive/*N* total).

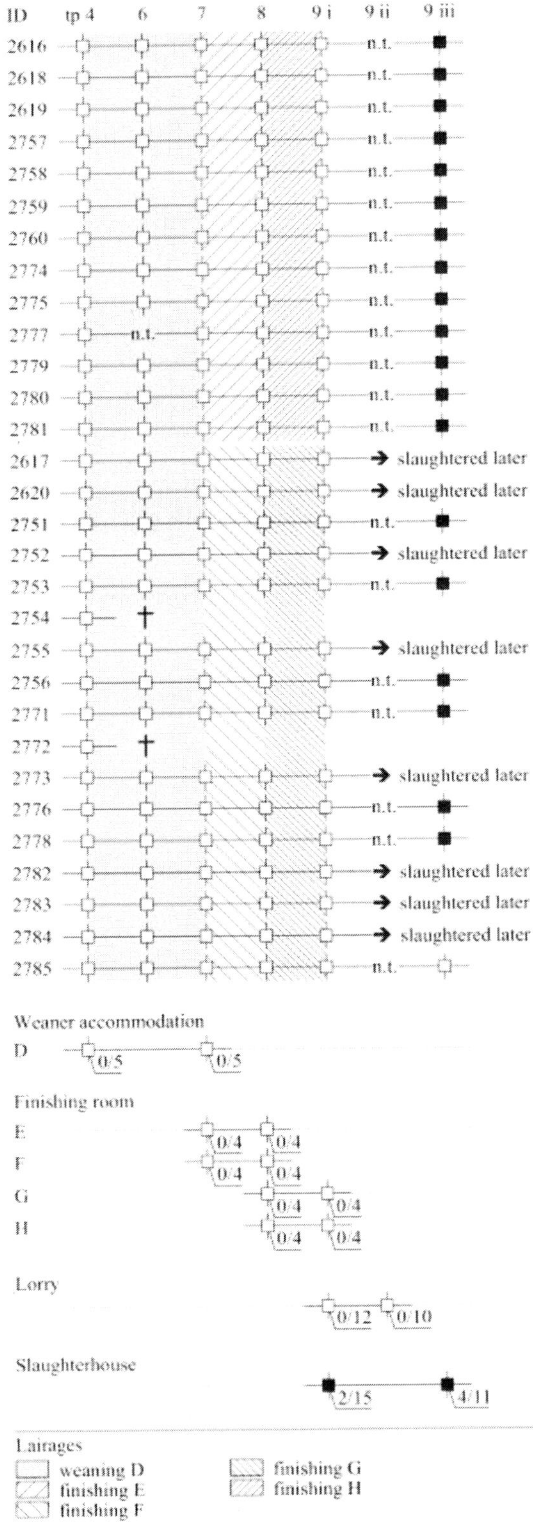
ID
tp 4
6
7
8
9 i
9 ii
9 iii
2616
2618
2619
2757
2758
2759
2760
2774
2775
2777
2779
2780
2781
2617
2620
2751
2752
2753
2754
2755
2756
2771
2772
2773
2776
2778
2782
2783
2784
2785
n.t.
slaughtered later
Weaner accommodation
D
0/5
Finishing room
E
F
G
H
0/4
Lorry
0/12
0/10
Slaughterhouse
2/15
4/11
Lairages
weaning D
finishing E
finishing F
finishing G
finishing H

At the end of finishing, the MRSA status of the individual pigs was intermittent during the entire fattening cycle (Fig. 2). The prevalence at tp8 and 9i was 7.9% (3/38) and 36.8% (14/38), respectively, and therefore was much lower than at earlier time points. In parallel, high percentage changes (94.1%) from a MRSA positive to MRSA negative status was observed from tp7 to tp8 (Table 3). Even after transportation (tp9ii) and after stunning (tp9iii), the individual MRSA status of a pig could have changed (Fig. 2). Only seven pigs (35.0%; 7/20) that tested MRSA negative at tp9i and 9ii remained negative even after stunning (tp9iii). Ten supplemental pigs were finished and slaughtered together with the study pigs but only tested at the end of fattening at tp9 immediately before transportation (Fig. 2). Only one pig was MRSA positive at tp9i; however, another four were positive at tp9iii, and two had changed from MRSA positive after transportation (tp9ii) to negative after stunning (tp9iii). Slaughter pigs (n = 42) from one other farm were transported on the same lorry to the abattoir and also tested at tp9 at the end of fattening (data not shown). In three of these animals, the intermittent MRSA status can also be seen, i.e., changing from MRSA negative at tp9i and 9ii to positive after stunning (tp9iii). Environmental wipes taken from the lorry previous to transportation were all MRSA negative, whereas nearly all samples after transport of the pigs were MRSA positive. Environmental wipes taken from the unused lairages at slaughterhouse were all MRSA negative (Fig. 2).

On **farm gf-I**, thirty-seven grower pigs out of three weaner groups (K, L and M) from farm ff-I were purchased with approximately eight weeks and penned together for fattening (Fig. 3). Sixteen pigs were tested until tp9iii, whereas another 21 pigs were only tested at tp7 and tp8, because they were slaughtered later. The MRSA status of individual pigs was intermittent during the entire fattening cycle. This also can be shown by the percentage changes of the MRSA status of the pigs, which were comparable to the data of farm ff-I (Table 3). In general the prevalence was decreasing, i.e., 50.0% (18/36) at tp8 and 25% (4/16) at tp9i. Environmental wipes from the finishing pen were MRSA positive previous to penning (tp7), as well as in the middle of the fattening cycle (tp8).

Even after transportation (tp9ii) and after stunning (tp9iii), the MRSA status of individual pigs had changed (Fig. 3). Five supplemental pigs were finished at farm gf-I and slaughtered together with the study pigs but only tested at the end of fattening at tp9. All pigs were MRSA negative at tp9i and 9ii. Interestingly, two pigs became MRSA positive at tp9iii after stunning. Environmental wipes taken from the unused lairages at slaughterhouse were all MRSA negative (Fig. 3).

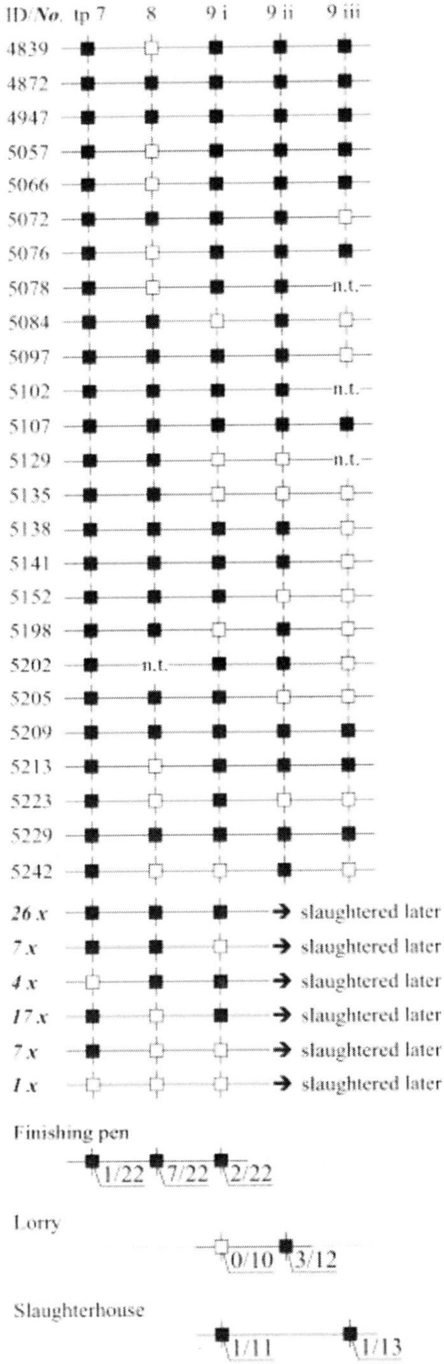

Figure 5. Fattening pigs on farm gf-II. □, MRSA-negative; ■, MRSA-positive.

On **farm ff-II**, all three sows (sows A, B and C) sampled before farrowing at tp1 were MRSA negative, as were the environmental wipes from the farrowing pens prior to holding pigs (data not shown). None of the three sows changed their MRSA status until the piglets were weaned (tp4) at approximately 35 days of age. At 24 h after farrowing (tp2), the offspring (n = 39) were completely negative. Also, the environmental wipes from the farrowing pens after holding pigs were MRSA negative (Fig. 4).

All weaned pigs stayed MRSA negative at all tp sampled (tp4, 6 and 7) throughout weaning. The MRSA status was consistently MRSA negative until tp9i, but changed to MRSA positive after stunning (tp9iii) in all but one animal. The percentage changes from a MRSA negative to MRSA positive status at tp9ii to tp9iii was thereby 95% (Table 3). Environmental wipes from the pens at the finishing room and from the lorry after transportation were consistently MRSA negative. In contrast environmental wipes from the lairages at the slaughterhouse were MRSA positive before, as well as after, holding pigs (Fig. 4).

On **farm gf-II**, eighty-seven grower pigs were merged together at approximately ten weeks (tp7) in one group and used for fattening (Fig. 5). The MRSA status of individual pigs was intermittent during the entire fattening cycle. In general, the prevalence decreased at the beginning of fattening (tp7 and 8) but slightly increased towards the end of fattening (tp9i), i.e., 94.3% (82/87), 61.6% (53/86) and 77.0% (67/87), respectively. Environmental wipes from the finishing pen were MRSA positive, before penning, as well as after holding pigs (Fig. 5). Also after transportation (tp9ii) and after stunning (tp9iii), the MRSA status of pigs had changed (Fig. 5). Fig. 6 shows results from additional slaughter pigs from two other farms; these pigs were transported on the same lorry as the study pigs from farm gf-II to the abattoir. In the majority of the pigs, intermittent MRSA status can be seen (Fig. 6). The overall dynamics of percentage changes of the MRSA status was shown to be comparable to farm ff-I and farm gf-I (Table 3). Environmental wipes taken from the lorry prior to transporting the pigs were all MRSA negative, whereas nearly one-third of the samples after transporting the pigs were MRSA positive. Environmental wipes from the lairages at the slaughterhouse were MRSA positive before as well as after holding pigs (Fig. 5).

4. DISCUSSION

The present study investigated the transmission of MRSA among individual pigs within a herd from birth to slaughter and between herds at transportation and slaughter. A better understanding of the critical points that are potentially responsible for both the spread and persistence of MRSA has been achieved.

The majority of all pigs from a MRSA-positive herd changed the MRSA status throughout the entire production cycle at least two times (Fig. 1, Fig. 2 and Fig. 3 and Fig. 5). The prevalence varied widely from 7.9% to 94.3% throughout the observation period. Varying prevalence has also been demonstrated in various other studies (Burns et al., 2014,Dewaele et al., 2011 and Weese et al., 2011).

At farm ff-I the highest percentage change from a MRSA negative to MRSA positive status was seen within piglets less than fourteen days of age, whereas changes from MRSA positive to MRSA negative were highest at the beginning of the fattening period (Table 3). Intermittent MRSA colonisation in individuals, as shown in the current survey, implies that pigs are transiently rather than permanently colonised. Little is known about the mechanisms that are responsible for such conditions. In humans, persistent colonisation occurs only in 20% of cases, 60% are intermittent carriers, and 20% are non-carriers (Wertheim et al., 2005 and Williams, 1963). Persistent carriers are frequently colonised by a single strain of *S. aureus,* and the load is higher (Eriksen et al., 1995,Nouwen et al., 2005, Nouwen et al., 2004b and VandenBergh et al., 1999). The MRSA status in farmers versus humans without regular contact to livestock has been investigated in other studies (Köck et al., 2012 and Van Cleef et al., 2011). The results ofVan Cleef et al. (2011) indicated that short-term exposure of humans on MRSA-positive farms frequently results in the acquisition of MRSA, but the majority of persons lose the MRSA strain within 24 h. They stated that the high prevalence of MRSA carriage in farmers and veterinarians could partly be the result of repeated contamination instead of real persistent colonisation. Köck et al. (2012) showed that 45.7% of farmers were persistently colonised with LA-MRSA, even over periods of non exposure. But 54.3% of the farmers turned out to be non-carriers or intermittent carriers. This situation may also be true for pigs, as the intermittent colonisation might point to a repeated contamination in a MRSA-positive environment as described above. Host characteristics and an optimal balance between the forces of defence and attraction seem to be responsible for the *S. aureus* carrier state (Nouwen et al., 2004a). Further studies are needed to investigate the underlying factors of MRSA contamination versus colonisation in pigs. Two time points are of major interest for such further investigations; piglets, which changed frequently from a MRSA negative to MRSA positive status within the first fourteen days of life and on the other hand grower pigs, often became MRSA negative at the beginning of the fattening period.

Transportation and processes at the slaughterhouse turned out to be critical control points for the spread of MRSA. At farm ff-II all pigs tested MRSA negative throughout the entire production cycle, but tested MRSA positive after stunning (Fig. 4). We also observed that pigs from another farm, tested MRSA negative prior to transport, but tested MRSA positive after transportation on a lorry with MRSA-positive pigs (Fig. 6). These findings are in agreement with other studies (Broens et al., 2011 and de Neeling et al., 2007). The most probable sources of MRSA are lorries, lairages and/or contact to other pigs, either transported together on the same lorry or in contact at slaughterhouse. Our findings support this assumption, as environmental wipes taken from lairages at slaughterhouse sh-II were MRSA positive before and after MRSA-negative pigs were housed there. As a consequence, all but one pig (19/20) that were held in this MRSA-positive lairage changed MRSA status from negative to positive after stunning. In the same slaughterhouse, the probability of MRSA transmission from the environment to the pigs is further increased by the fact that pigs are kept at least two hours at the lairage prior to slaughter. To avoid the transmission of MRSA from the lairages to slaughter pigs, a strict hygiene management is required, such as cleaning lairages for every new slaughter batch and avoiding long resting times in lairages before slaughtering. Successful cleaning is feasible since environmental wipes from other slaughterhouses were tested MRSA negative in our study. Possibly, MRSA transmission from pig to pig may be minimised if contact between pigs on lorries or at slaughterhouse is strictly impossible, e.g., using completely closed fences. Another possibility could be to transport only MRSA-negative pigs instead of mixing pigs with unknown MRSA status.

Figure 6. Slaughter pigs of two other farms on the transport batch gf-II, sampled at tp 9 only. □, MRSA-negative; ■, MRSA-positive. ID, ear tag; No., number of pigs with the same result; tp, time point; n.t., not tested; n.e., no ear tag; 2/15, environmental samples (*n* positive/*N* total).

	ID/*No.*	tp 7	8	9 i	9 ii	9 iii
farm 1	2126	n.t.	n.t.	□	□	■
	2172	n.t.	n.t.	□	□	■
	2196	n.t.	n.t.	□	■	□
	2245	n.t.	n.t.	■	□	□
	2535	n.t.	n.t.	■	■	■
	2545	n.t.	n.t.	□	n.t.	■
	2557	n.t.	n.t.	□	□	n.t.
	2581	n.t.	n.t.	■	■	□
	2585	n.t.	n.t.	□	■	■
	2600	n.t.	n.t.	□	□	■
	2912	n.t.	n.t.	■	□	■
	2913	n.t.	n.t.	■	■	■
	2914	n.t.	n.t.	□	■	■
	4293	n.t.	n.t.	■	■	■
	4322	n.t.	n.t.	■	□	■
	4325	n.t.	n.t.	□	■	□
	4329	n.t.	n.t.	□	n.t.	■
	4399	n.t.	n.t.	□	■	■
	4406	n.t.	n.t.	■	■	n.t.
	4409	n.t.	n.t.	□	□	■
	7222	n.t.	n.t.	□	■	■
	7250	n.t.	n.t.	□	□	■
	7264	n.t.	n.t.	□	□	■
	7266	n.t.	n.t.	□	■	□
	7269	n.t.	n.t.	□	■	n.t.
	7280	n.t.	n.t.	□	■	□
	7282	n.t.	n.t.	□	□	■
	7284	n.t.	n.t.	□	□	■
	7310	n.t.	n.t.	□	■	■
	7338	n.t.	n.t.	□	□	n.t.
	7343	n.t.	n.t.	□	□	n.t.
	7344	n.t.	n.t.	□	■	■
	7353	n.t.	n.t.	□	■	□
	7358	n.t.	n.t.	□	□	■
	7368	n.t.	n.t.	□	□	■
	n.e.	n.t.	n.t.	□	□	■
	20 x	n.t.	n.t.	□	□	□
farm 2	3092	n.t.	n.t.	□	□	■
	3190	n.t.	n.t.	□	□	n.t.
	3193	n.t.	n.t.	□	■	□
	3196	n.t.	n.t.	□	□	■
	3226	n.t.	n.t.	□	□	■
	3233	n.t.	n.t.	□	□	■
	3254	n.t.	n.t.	□	□	■
	14 x	n.t.	n.t.	□	□	□

Table 3. Rates of changes of the MRSA status of the pigs for every transition from one time point to another.

	tp2–3			tp3–4			tp4–6[a]			tp6–7		
	% n–p	% p–n	% n.ch.	% n–p	% p–n	% n.ch.	% n–p	% p–n	% n.ch.	% n–p	% p–n	% n.ch.
Farm ff-I (95% CI)	80.3 (68.2–89.4)	10.2 (2.9–24.2)	47.0 (36.9–57.2)	37.5 (1.5–64.6)	24.1 (15.1–35.0)	73.7 (63.6–82.2)	63.6 (30.8–89.1)	21.4 (8.3–41.0)	66.7 (49.8–80.9)	33.3 (7.5–70.1)	46.4 (27.5–66.1)	56.8 (39.5–72.9)
Farm ff-II (95% CI)	0.0 (0.0–11.2)	n.c[b]	100.0 (88.8–100.0)	0.0 (0.0–11.6)	n.c	100 (88.4–100.0)	0.0 (0.0–12.8)	n.c	100.0 (87.2–100.0)	0.0 (0.0–12.8)	n.c	100.0 (87.2–100.0)
Farm gf-I (95% CI)	–[c]	–	–	–	–	–	90.9 (55.5–99.8)	14.8 (4.2–33.7)	64.9 (47.5–79.8)	100.0 (4.8–100.0)	12.5 (3.5–29.0)	75.7 (58.8–88.2)

	tp7–8			tp8–9i			tp9i–9ii			tp9ii–9iii		
	% n–p	% p–n	% n.ch.	% n–p	% p–n	% n.ch.	% n–p	% p–n	% n.ch.	% n–p	% p–n	% n.ch.
Farm ff-I (95% CI)	5.3 (0.1–26.0)	94.1 (71.3–99.9)	50 (32.9–67.1)	41.2 (24.6–59.3)	100 (29.2–100.0)	54.1 (36.9–70.5)	26.7 (7.8–55.1)	71.4 (29.0–96.3)	59.1 (36.4–79.3)	33.3 (11.8–61.6)	100 (47.8–100.0)	50 (27.2–72.8)
Farm ff-II (95% CI)	0.0 (0.0–12.3)	n.c	100.0 (87.7–100.0)	0.0 (0.0–12.3)	n.c	100.0 (87.7–100.0)	n.t.	n.t.	n.t.	95.0[d] (75.1–99.9)	n.c.	5.0[d] (0.1–24.9)
Farm gf-I (95% CI)	50 (6.8–93.2)	50 (31.9–68.1)	50 (32.9–67.0)	12.5 (3.2–52.7)	57.1 (18.4–90.1)	66.7 (38.4–88.2)	0.0 (0.0–28.5)	100.0 (39.8–100.0)	73.3 (44.9–92.2)	46.7 (21.2–73.4)	n.c.	53.3
Farm gf-II (95% CI)	80.0 (28.4–99.5)	39.5 (28.8–51.0)	58.1 (47.0–68.7)	72.7 (54.5–86.7)	20.8 (10.8–34.1)	59.3 (48.2–69.8)	60 (14.7–94.7)	15 (3.2–37.9)	76.0 (54.9–90.6)	0.0 (0.0–60.2)	44.4 (21.5–69.2)	63.4 (40.7–82.8)

tp: time point, n–p: change from MRSA negative to positive, p–n: change from MRSA positive to negative, n.ch.: no change, n.t.: not tested, CI: confidence intervals.

[a] Time point 5 has been omitted because only environmental wipes were taken.

[b] n. c.: not calculated, changes cannot be calculated as there are no positive pigs.

[c] At time point 2, 3, 4 and 6, the piglets/weaned pigs were still housed on farm ff-I. Results were included in the results from farm ff-I. From transition tp4 to tp6 on, results were displayed at farm gf-I.

[d] At time point 9ii, it was not possible taking nasal swabs at the slaughterhouse, therefore the results refer to the transition from time point 9i to 9iii.

The necessity for defining a MRSA status at the herd level instead of the level at the individual pig is clearly demonstrated in our study. The determination of the MRSA status of an individual pig reflects only a short moment in life. Because this status can change immediately, the separation within a herd of MRSA-positive from MRSA-negative pigs as a basis for subsequent eradication is not possible. In contrast, defining MRSA-positive and negative farms by screening batches of individual pigs and/or the environment was shown to be very reliable. Therefore, defining farms either as MRSA positive or negative and limiting animal trade to herds of the same status is highly recommended, when the spread of MRSA should be prevented. Our results demonstrated that it is essential for MRSA-negative farms to purchase pigs only from breeders that are certified as MRSA-free.

The determination of the MRSA prevalence in Swiss pig herds based on single pig testing at slaughter, which leads to an imprecise estimation of prevalence at the farm level. However, sampling at farm level is much more time-consuming and costly. Alternatively, defining MRSA status at herd level, followed by subsequent control measures during the entire production cycle, could lead to a more accurate determination of MRSA prevalence, even with sampling at the slaughterhouse. Additionally, it should also be envisaged to examine the status of methicillin-sensitive *Staphylococcus aureus* (MSSA) in pig farming which may also be resistant to multiple antibiotics and possibly harbour virulence factors like Panton-Valentine leukocidin or other toxins.

5. CONCLUSIONS

Intermittent colonisation throughout the entire production cycle was conspicuous in the tracking of MRSA at the level of the individual pig. This implies that pigs are transiently rather than permanently colonised and suggests repeated contamination. As a consequence, the MRSA status should be defined at the herd level instead of individual pigs. As the prevalence in Swiss slaughter pigs is constantly increasing, the further spread of MRSA could be prevented by defining farms as MRSA positive or negative and allowing the trade of animals only within herds of the same status. With the implementation of these measures, also further dissemination of LA-MRSA into healthcare facilities and the community via humans with close contact to these animals, i.e., farmers, veterinarians and slaughterhouse workers could be prevented.

ACKNOWLEDGEMENTS

We are grateful to all farmers and staff who granted access to their farms and their kind cooperation; to the slaughterhouse managers and their team, who provide the access to the slaughterhouses and the organisation of taking samples from animals and their facility; to the lorry-drivers for their patience and their permission taking samples of the vehicle; to Patricia Hirsiger, Bettina Jenny, Julia Malik, Dolf Kümmerlen and Wolfgang Pendl, Vetsuisse Faculty, University of Zurich, Switzerland, Saria In-Albon, Julia Furrer, Bettina Meuli, Claudio Dolder and Thomas Mock, Vetsuisse Faculty, University of Bern, Switzerland, for their assistance in taking samples on farms and slaughterhouse; and to Isabelle Brodard and Susanne Rickli, Vetsuisse Faculty, University of Bern, Switzerland, for excellent technical support. This study was financed by a research grant of the Swiss Federal Food Safety and Veterinary Office (FSVO), grant number 1.14.k.

REFERENCES

1. Broens, E.M., Espinosa-Gongora, C., Graat, E.A., Vendrig, N., van der Wolf, P.J., Guardabassi, L., Butaye, P., Nielsen, J.P., de Jong, M.C., van de Giessen, A.W., 2012. Longitudinal study on transmission of MRSA CC398 within pig herds. BMC Vet. Res. 8, 58.

2. Broens, E.M., Graat, E.A., van der Wolf, P.J., van de Giessen, A.W., de Jong, M.C., 2011. Transmission of methicillin resistant Staphylococcus aureus among pigs during transportation from farm to abattoir. Vet. J. 189, 302–305.

3. Burns, A., Shore, A.C., Brennan, G.I., Coleman, D.C., Egan, J., Fanning, S., Galligan, M.C., Gibbons, J.F., Gutierrez, M., Malhotra-Kumar, S., Markey, B.K., Sabirova, J.S., Wang, J., Leonard, F.C., 2014. A longitudinal study of Staphylococcus aureus colonization in pigs in Ireland. Vet. Microbiol. 174, 504–513.

4. Büttner, S., Mehmann, M., Müntener, C., Torriani, K., Overesch, G., 2014. Bericht über den Vertrieb von Antibiotika in der Veterinärmedizin und das Antibiotikaresistenzmonitoring bei Nutztieren in der Schweiz (ARCH-VET 2013) (http://www.blv.admin.ch).

5. Crombé, F., Argudin, M.A., Vanderhaeghen, W., Hermans, K., Haesebrouck, F., Butaye, P., 2013. Transmission dynamics of methicillin-resistant Staphylococcus aureus in pigs. Front. Microbiol. 4, 57.

6. Dewaele, I., Messens, W., De Man, I., Delputte, P., Herman, L., Butaye, P., Heyndrickx, M., Rasschaert, G., 2011. Sampling, prevalence and characterization of methicillin-resistant Staphylococcus aureus on two Belgian pig farms. Vet. Sci. Dev. 1.

7. de Neeling, A.J., van den Broek, M.J.M., Spalburg, E.C., van Santen-Verheuvel, M.G., Dam-Deisz, W.D.C., Boshuizen, H.C., van de Giessen, A.W., van Duijkeren, E., Huijsdens, X.W., 2007. High prevalence of methicillin resistant Staphylococcus aureus in pigs. Vet. Microbiol. 122, 366–372.

8. Eriksen, N.H., Espersen, F., Rosdahl, V., Jensen, K., 1995. Carriage of Staphylococcus aureus among 104 health persons during a 19-month period. Epidemiol. Infect. 115, 51–60.

9. Köck, R., Loth, B., Köksal, M., Schulte-Wülwer, J., Harlizius, J., Friedrich, A.W., 2012. Persistence of nasal colonization with livestock-associated methicillin-resistant Staphylococcus aureus in pig farmers after holidays from pig exposure. Appl. Environ. Microbiol. 78, 4046–4047.

10. Lewis, H.C., Mølbak, K., Reese, C., Aarestrup, F.M., Selchau, M., Sørum, M., Skov, R.L., 2008. Pigs as source of methicillin-resistant Staphylococcus aureus CC398 infections in humans, Denmark. Emerg. Infect. Dis. 14, 1383–1389.

11. Nouwen, J., Boelens, H., van Belkum, A., Verbrugh, H., 2004a. Human factor in Staphylococcus aureus nasal carriage. Infect. Immun. 72, 6685–6688.

12. Nouwen, J.L., Ott, A., Kluytmans-Vandenbergh, M.F., Boelens, H.A., Hofman, A., van Belkum, A., Verbrugh, H.A., 2004b. Predicting the Staphylococcus aureus nasal carrier state: derivation and validation of a culture rule. Clin. Infect. Dis. 39, 806–811.

13. Nouwen, J.L., Fieren, M.W., Snijders, S., Verbrugh, H.A., van Belkum, A., 2005. Persistent (not intermittent) nasal carriage of Staphylococcus aureus is the determinant of CPD-related infections. Kidney Int. 67, 1084–1092.

14. Overesch, G., Büttner, S., Perreten, V., 2012. Evolution of methicillin-resistance Staphylococcus aureus (MRSA). Fleischwirtsch. Int. 6, 61–63.

15. Stegger, M., Lindsay, J.A., Moodley, A., Skov, R., Broens, E.M., Guardabassi, L., 2011. Rapid PCR detection of Staphylococcus aureus clonal complex 398 by targeting the restriction-modification system carrying sau1-hsdS1. J. Clin. Microbiol. 49, 732–734.

16. Van Cleef, B., Graveland, H., Haenen, A., Van de Giessen, A., Heederik, D., Wagenaar, J., Kluytmans, J., 2011. Persistance of livestock-associated methicillin-resistant Staphylococcus aureus in field workers after short-term occupation exposure to pigs and veal calves. J. Clin. Microbiol. 49, 1030–1033.

17. van Rijen, M.M.L., van Keulen, P.H., Kluytmans, J.A., 2008. Increase in a Dutch Hospital of methicillin-resistant Staphylococcus aureus related to animal farming. Clin. Infect. Dis. 46, 261–263.

18. VandenBergh, M.F., Yzerman, E.P., van Belkum, A., Boelens, H.A., Sijmons, M., Verbrugh, H.A., 1999. Follow-up of Staphylococcus aureus nasal carriage after 8 years: redefining the persistent carrier state. J. Clin. Microbiol. 37, 3133–3140.

19. Verhegghe, M., Pletinckx, L.J., Crombe, F., Van Weyenberg, S., Haesebrouck, F., Butaye, P., Heyndrickx, M., Rasschaert, G., 2013. Cohort study for the presence of livestock-associated MRSA in piglets: effect of sow status at farrowing and determination of the piglet colonization age. Vet. Microbiol. 162, 679–686.

20. Verkade, E., Kluytmans, J., 2014. Livestock-associated Staphylococcus aureus CC398: animal reservoirs and human infections. Infect. Genet. Evol. 21, 523–530.

21. Voss, A., Loeffen, F., Bakker, J., Klaassen, C., Wulf, M., 2005. Methicillin-resistant Staphylococcus aureus in pig farming. Emerg. Infect. Dis. 11, 1965–1966.

22. Weese, J.S., Zwambag, A., Rosendal, T., Reid-Smith, R., Friendship, R., 2011. Longitudinal investigation of methicillin-resistant Staphylococcus aureus in piglets. Zoonoses Public Health 58, 238–243.

23. Wertheim, H.F.L., Melles, D.C., Vos, M.C., van Leeuwen, W., van Belkum, A., Verbrugh, H.A., Nouwen, J.L., 2005. The role of nasal carriage in Staphylococcus aureus infections. Lancet. Infect. Dis. 5, 751–762.

24. Wettstein Rosenkranz, K., Rothenanger, E., Brodard, I., Collaud, A., Overesch, G., Bigler, B., Marschall, J., Perreten, V., 2014. Nasal carriage of methicillin-resistant Staphyloccocus aureus (MRSA) among Swiss veterinary health care providers: Detection of livestock- and healthcare-associated clones. 156, 317–325.

25. Williams, R.E., 1963. Healthy carriage of Staphylococcus aureus: its prevalence and importance. Bacteriol. Rev. 27, 56–71.

CHAPTER 9

MRSA Epidemiology in Animals

Patrícia Yoshida Faccioli-Martins and Maria de Lourdes Ribeiro de Souza da Cunha

UNESP- Univ. Estadual Paulista, Biosciences Institute, Botucatu, SP Brazil

1. INTRODUCTION

Until the 1990s, methicillin-resistant Staphylococcus aureus (MRSA) was traditionally considered a pathogen causing nosocomial infections, being the so-called HA-MRSA (healthcare-associated methicillin-resistant Staphylococcus aureus). However, over time, cases of MRSA-positive individuals were observed who never had contact with hospital services, and strains from these individuals were identified and named CA-MRSA (communityassociated methicillin-resistant Staphylococcus aureus). In 2003 in the Netherlands, a new MRSA strain arose in patients that could not be typed through PFGE (pulsed field gel electrophoresis) with SmaI, with resistance to digestion by this enzyme (Bens et al., 2006), being called since then NT-MRSA (non typeable methicillin-resistant Staphylococcus aureus). Investigations of this NT-MRSA intensified, and it was observed that these patients carrying this strain had previous contact with pigs and the geographic distribution of cases showed clusters near pig farms (van Loo et al., 2007). With more advanced studies, it was possible to determine strains strictly related to animals, such as those found in pigs, which were named LA-MRSA (livestock-associated methicillin-resistant Staphylococcus aureus) in 2010 (Vanderhaeghen et al., 2010).

The resistance to methicillin in staphylococci is mediated by the mecA gene that encodes a modified penicillin-binding protein (PBP), the

PBP2a or 2', which shows reduced affinity to the resistant penicillins to penicillinase, such as methicillin and oxacillin and for all other beta-lactam antibiotics (van Duijkeren et al., 2004). Due to the need for better characterizing these isolates, they have been classified in a more detailed manner, beginning with the SCCmec types and patterns identified by PFGE, and are currently based on sequence type and spa typing. With the use of techniques such as MLST (multi locus sequence typing) and spa typing, characteristic clones from animals have been observed, and it is suspected that some have tropism by determined host species. An example of this is ST398, which in addition to being strictly linked to pigs, has carried novel types of SCCmec. Li et al. (2011) analyzed the SCCmec element structure carried by 31 CC398 MRSA strains isolated from participants of a conference. The strains were classified into novel types, IX and X, type V (5C2&5) subtype c and type IVa, all carriers of genes conferring resistance to metals. The SCCmec structures from CC398 strains were distinct from those usually found in humans, complementing evidence that humans are not the original host for CC398. With the absence of a complete evaluation of risk factors for carrying this strain as much in animals as in humans, Graveland et al. (2010) carried out the first study that showed direct association between animal and human carriage of ST398. This association, in addition to the association between MRSA and the antimicrobial use in calves, highlights the need for prudent use of antibiotics in farm animals.

One clone of special interest in ruminants is the CC133. The great majority of isolates from small ruminants is represented by this clonal complex of S. aureus, but its evolutionary origin and molecular basis for its host tropism remain unknown. Guinane et al. (2010), attempting to determine whether the CC133 developed as result of a transmission from human host to ruminant followed by an adaptative diversification of the genome, carried out a comparative sequencing of the complete genome. Several novel mobile genetic elements were observed in the CC133 isolates encoding virulence proteins with attenuated or enhanced activity and they were widely distributed, suggesting a key role in their hostspecific interaction. These data provide broad and new insights into the origin and basis molecular of S. aureus ruminant host specificity. The MRSA evolution and epidemiology in animals are discussed in this chapter.

2. HISTORY OF MRSA IN ANIMALS

The first report of MRSA in animals was in milk from Belgium cows with mastitis (Morgan, 2008). Until 2000, MRSA had been isolated sporadically from animals, in particular cows, small companion animals, and horses. With exception of some equine isolates, the nature of these cases suggested a human origin and no epidemics have been reported. In this respect, until the end of 20th century, both the scientific community and policy makers were convinced that animal husbandry was of little relevance for MRSA causing diseases in humans, but was particularly a problem based on antimicrobial use in human medicine. The situation has changed with a growing number of reports of MRSA in livestock, especially pigs and veal calves. MRSA has also been reported in companion animals and horses, as well as transmission between humans and animals (Catry et al., 2010). Calling attention to this dramatic increase of MRSA in animals, van Duijkeren et al. (2010) at the Veterinary Microbiological Diagnostic Center, in the Netherlands, reported 0% MRSA in isolates from equine clinical samples in 2002 and then 37% in 2008.

3. THE IMPORTANCE OF THE DIFFERENT ANIMAL SPECIES

Various species have been identified as host and carriers of MRSA in different countries and settings, including dogs, cats, sheep, chickens, horses, pigs, rabbits, seals, psittacine birds, and one turtle, bat, guinea pig and chinchilla (Morgan, 2008). With differences between strains isolated from pets, wild animals and cattle, it is important to evaluate each species individually, because they present peculiar characteristics, including the series of resistance genes. Thus, the concern for MRSA types in animals has grown considering their role as a potential reservoir or vector for human infection by MRSA in the community. However the data available on MRSA transmission between humans and companion animals are limited, and further epidemiological studies are needed on this transmission and its impact on public health (Loeffler & Lloyd, 2010)

3.1 Companion animals

Pets have been shown to act as reservoirs of bacteria resistant to antimicrobials, and MRSA transmission between humans and animals has been described. For strains of MRSA with a low specificity to the host, the transference is likely to occur in both directions between humans and pets living at the same household (Nienhoff et al., 2009). The infections by MRSA in companion animals are predominantly of skin and soft tissues, especially during post-surgical (Morgan, 2008).

Nienhoff et al. (2009) reported two cases of transmission of MRSA strains between humans and dogs. Three positive dogs to MRSA were identified in a survey carried out in 803 dogs and 117 cats admitted to the Small Animal Clinic of the University of Veterinary Medicine Hannover, Germany. The first case was a 6-month-old female admitted to the clinic for teeth extraction. The owner, MRSA-positive, was a specialist veterinarian in swine diseases, working in pig barns 4-5 days per week and having access to MRSA ST398-positive farms. The dog and owner strains were identical through molecular typing, belonging to ST398 and spa type t034. The second case was an 11-year-old male admitted to the clinic because of a cardiac problem. The likely origin of the strain was the mother-in-law of the dog's owner, who is diabetic, having received nursing care at home and presenting an infected wound on the foot and an ulcer in the right eye. The MRSA isolates found in these lesions and in the dog belonged to ST225, which is frequently found in humans.

3.2 Cattle

Cows with mastitis have been the most likely to harbor MRSA, and they may be related to horizontal transfer via wet hands of colonized or infected dairy farm workers, and selection by the use of antibiotics to treat mastitis (Morgan, 2008). The first known case of MRSA transmission between cows and a person was reported by Juhász-Kaszanyitzky et al. (2007). MRSA strains isolated from cows with subclinical mastitis were phenotypically and genotypically indistinguishable from the strain from the person who worked with these animals. These strains were determined as ST1, spa type t127, SCCmecIVa. The authors considered these strains epidemiologically related, indicating transmission from cow to human or from human to cow.

Feβler et al. (2010) studied 25 MRSA ST398 isolates from cases of bovine clinical mastitis and two isolates from farm workers originating from 17 dairy farms in Germany, evaluating the genetic relatedness, antimicrobial resistance and virulence properties. Nine major ApaI PFGE patterns were found, three spa types (t011, t034 and t2576) and two types of SCCmec (IV and V) were identified. As described previously for ST398 from pigs, isolates from this sequence type originating in cases of bovine mastitis have also shown a high degree of variability when the ApaI PFGE profile and other genotypic and phenotypic characteristics were compared. A uniform pattern of virulence genes appeared to be conserved between ST398 isolated from both animal species.

Türkyılmaz et al. (2010) detected 14 of 16 strains from bovine milk of the lineage MRSA ST- 239-III in which one was related to hospital-associated clones, and two strains were ST8/IV, which correspond to USA300, which causes severe community-acquired infections. The presence of MRSA ST239-III lineage can indicate a transmission from humans to animals, and the presence of ST8-IV can show emergence of strains from the community in the Aydin region in Turkey. This study underscores the necessity to take measures to avoid MRSA transmission between humans and animals.

3.3 Horses

In horses, MRSA have been reported in infections of skin and soft tissues, bacteraemia, septic arthritis, osteomyelitis, implant-related infections, metritis, omphalitis, catheterrelated infections and pneumonia. The first MRSA outbreak in horses was observed in 1993, with 11 infected horses in the post-surgical in a veterinary teaching hospital in Michigan. Subsequent outbreaks have occurred on Japan, Austria, the UK, Ireland, Canada, and other areas of the USA (Morgan, 2008).

In 2009, Loeffler et al. reported the first isolation of MRSA ST398 spa type t011 in animals from the UK. They were two horses in southeastern England, with isolates with identical phenotypic and genotypic characteristics as reported in horses in Belgium, Austria and Germany, which also carried the SCCmec type IVa. They vary from those commonly found in pigs (spa type t108, t034 or t571) and frequently carry SCCmec V, possibly indicating hostspecific variation within this lineage or independent evolution. One interesting fact is that isolates from pigs and horses commonly show resistance to tetracycline and/or gentamicin, both agents frequently used in pigs and horses, respectively.

These findings demonstrate the introduction of ST398 in England and provide more evidence of successful dissemination of this zoonotic pathogen in the animal reservoir. The authors recommended vigilance for MRSA ST398 in both animals and humans.

Outbreaks of MRSA were observed in horses and horse personnel in the Netherlands in the period of 2006-2008. The isolates belonged to ST8, spa type t064, and to ST398, spa types t011 and t2123, predominantly. During the outbreak of post-surgical infections by MRSA in horses in a veterinary teaching hospital, isolates from spa type t2123 were isolated from 7 horses, and 4 of 61 personnel indicating a zoonotic transmission; after intervention, the outbreak stopped. In another outbreak that occurred in 2008, 17 horses with MRSA were detected, with 12 spa type t011, 4 spa type t2123 and 1 spa type t064. From 170 personnel, 16 were positive for MRSA, with 11 spa type t011 and 5 spa type t2123. From 106 personnel who maintained close contact with horses, 15 were MRSA-positive compared with 1 MRSA-positive of 64 personnel who had no close contact with the animals. Furthermore, screening carried out on the horses on admission showed that 9.3% were MRSA-positive, predominantly spa-type t011. Weekly crosssectional sampling from all horses hospitalized for 5 weeks demonstrated that 42% were MRSA-positive at least once, again predominantly with spa type t011, which suggests that a nosocomial transmission appeared. The research of environmental samples from veterinary hospital revealed the presence of 53% of MRSA, including samples from students and staff member rooms, all spa type t011, indicating that humans contribute to the microorganism dissemination. The samples cultured employing pre-enrichment with high-salt concentration presented better results than the method without pre-enrichment. These results demonstrate that the nosocomial transmission in equine clinics occurs and suggest that personnel play a role in the transmission (van Duijkeren et al., 2010).

3.4 Poultry

The first reports of MRSA in chicken meat occurred in Korea in 2003 (Lee, 2003) and Japan in 2005 (Kitai et al., 2005), but it was not determined as livestock-associated, raising the hypothesis of meat contamination with human strains through handlers.

LA-MRSA was reported primarily in health poultry in 2008 in Belgium, with ten recent isolates classified as ST398 spa types t011 and t567. In this study, strains isolated in 1970s and strains isolated recently in 2006 were evaluated. It was observed that from 12 antimicrobial agents tested, eight presented percentage of resistance significantly higher in the recent isolates (Nemati et al., 2008).

Persoons et al. (2009) evaluated samples from 50 laying hens and 75 broiler chickens in Belgium. MRSA was found in 8 broiler chickens from 2 of 14 farms sampled, belonging to ST398 spa type t1456. According to the author, it still remains unclear as to whether this strain is associated with poultry.

3.5 Pigs

Mounting evidence suggests that livestock, particularly pigs, can represent an important reservoir of CA-MRSA (community-associated – CA) strains that can colonize and infect humans in close contact with them. ST398 is the most commonly reported MRSA sequence type among large livestock in Europe. These strains frequently carry genes encoding for non-beta-lactam antimicrobial resistance, including a plasmid-borne gene with resistance to trimetoprim, dfrK, identified in an isolate from a pig from Germany. Furthermore, these isolates are referred to often as nontypeable by PFGE because their genome is not digested by SmaI enzymes, several common spa types have been associated with them, carrying SCCmec types IV and V and they typically lack PVL genes (David & Daum, 2010).

In 2008, with the aim of evaluating whether other professionals in contact with pigs, in addition to farmers and veterinarians, have a higher risk of carrying MRSA than the population in general, a study was carried out with 272 participants of a conference on swine health in Denmark. In total, 34 (12.5%) participants from 9 countries carried MRSA, being that 31 isolates were not typeable by PFGE with SmaI. They belonged to ST398, spa types t011, t034, t108, t571, t567 and t899. The MRSA transmission from pigs to staff demonstrated to be an international problem, creating a new reservoir for a strain that was still considered CA-MRSA (Wulf et al., 2008).

Horgan et al. (2010) evaluated the prevalence of MRSA in the swine population in Ireland. A total of 440 pigs were evaluated from 41 geographically distributed farms and 100 individuals involved in the pig industry. No MRSA isolate was recovered from pigs but two humans tested were identified as MRSA carriers. These individuals were working in the wider pig industry. These isolates belonged to ST22 and ST1307, indicating that during the period of study the porcine colonization by MRSA, in particular the animal-related strain ST398, was not common in Ireland.

Wagenaar et al. (2009) published the first report of MRSA on pig farms in China, and it was the first time that MRSA ST9 in 4 pig farms and one single locus variant of MLST ST9 (ST1376) were detected in a pig farm. This study shows that LA-MRSA is not restricted to the clonal lineage ST398 found in Europe and North America in commercial pigs, but that other MRSA lineages are able to spread in the livestock in the same manner, also confirming that livestock can act as a reservoir of MRSA.

Despite that LA-MRSA appears be the predominant MRSA strain in pigs, some studies mention the detection of non-LA-MRSA strains in these animals, possibly by transmission of human strains to the animals. In Singapore, ST22-MRSA-IV was isolated from pigs and this strain was previously found increasingly important in the hospital population there. Notably the ST22-MRSA-IV is also a major hospital clone, as is the UK-EMRSA-15 found in the UK, with both strains indicating a contamination of human origin. In Canada, 14% of MRSA isolated from pigs appeared to belong to the human epidemic clone CMRSA-2 (Canadian epidemic MRSA-2, known as USA100), while 74.4% of isolates were LA-MRSA. The remaining strains belonged to rare clones, not related to LA-MRSA or CMRSA-2. Most reports on LA-MRSA in pigs originate from the Netherlands. In Europe, LA-MRSA has also been found in pigs in Germany, Denmark and Belgium, and beyond Europe, in Canada, Singapore and the USA (Vanderhaeghen et al., 2010).

3.6 Products of animal origin

Besides the importance of living animals as a source of MRSA, animal origin products also play a role in disseminating these strains to the humans. Lozano et al. (2009) detected MRSA ST398 in food samples in Spain. A total of 318 samples of raw food were evaluated from food-producing animals (148 from chicken, 55 from pork, 46 from veal, 19 from lamb, 10 from turkey, 8 from rabbit and 12 minced-meat samples) and of wild animals (8 game birds, 4 wild boar, 4 deer and 4 hare samples). MRSA was detected in 5 of 318 (1.6%) food samples (pork, chicken, rabbit, veal and wild boar). The two strains from pork and veal corresponded to ST398-SCCmecV clone (spa types t011 and t1197, respectively), the two strains from chicken and rabbit were typed as ST125-SCCmecIVa-t067, and the strain from one wild boar was ST217-

SCCmecIVa-t032, with all the MRSA being PVL negative. The characteristics of these strains suggest that they can be of both animal and human origin, and although the presence of MRSA in food is low, it must be monitored, because it can contribute to its dissemination

Recently, Weese et al. (2011) evaluated the presence of MRSA in feedlot cattle, close to the time of slaughter, in nasal and rectal fecal samples. It was not possible to detect MRSA in these animals, in contrast to recent studies on retail beef (Weese et al., 2010), demonstrating the need for more studies of livestock, as well as farms, processing and retail environments to elucidate the epidemiology of contamination with MRSA in meat.

4. MRSA AS ZOONOSIS OR HUMANOSIS

Several studies have been done to determine the degree to which MRSA plays a role in zoonosis or humanosis. It has been observed that usually the strains originating from companion animals are originally human strains, and that the infection with this MRSA type is considered humanosis. On the other hand, the strains orginating from livestock (livestockassociated - LA) are often divergent from human strains and the infection with this type of LA-MRSA could be considered zoonosis, and in this case MRSA would be an emergent zoonotic agent. Within this context, veterinarians, cattle farmers and pet owners are considered risk groups for acquiring MRSA (Morgan, 2008).

In 2011, a case of empyema was reported in a 79-year-old man in Spain with isolation of MRSA ST398 in purulent exudates from the thorax and trachea and nasal swabs. The ineffective initial therapy with levofloxacin was modified to intravenous linezolid, but the clinical situation of the patient rapidly deteriorated, and he died of multiorgan failure. The three MRSA strains were typed as ST398, spa-type t011, SCCmec V and agrI and presented the same phenotypic resistance, including β-lactam, tetracycline, clindamycin (but not erythromycin), ciprofloxacin and levofloxacin. It is important to point out that the patient lived with his wife and two sons near a pig farm. Both sons worked on the farm. The patient, but not his wife, sporadically helped on the farm. Nasal samples from three family members indicated MRSA carriage in one son. The characteristics of this isolate were identical to the isolate from the patient. Furthermore, nasal swabs of 18 pigs from the farm were randomly collected, and MRSA isolates were detected in 9 (50%) pigs.

One MRSA isolated in each animal was minutely characterized. Eight isolates were typed as ST 398/t011/SCCmec V/agrI and one remaining isolate as ST398/t1451/SCCmec V/agrI. All MRSA isolated from the animals had the same phenotypic and genotypic resistance comparing MRSA isolated from the patient and son. These findings indicate potential pighuman zoonosis transmission of MRSA ST398 and that this clone can be associated with severe respiratory pathology in immunocompromised patients, and this microorganism can also be resistant to other first-line antimicrobial agents, such as fluoroquinolones, used to treat these infections. Furthermore, the unusual clindamycin-resistance/erythromycinsusceptibility phenotype can be a key marker (in addition to tetracycline resistance) of the possible presence of livestock-associated MRSA (Lozano et al., 2011).

As seen in the reports, animal reservoirs for MRSA are becoming recognized worldwide with increasing awareness of MRSA ST398 colonizing pig and veal farmers, and attending veterinarians, at high rates. In The Netherlands, these groups are now considered high risk, and if admitted to a hospital, they are immediately conducted to isolation, screening for MRSA and decolonization (Loeffler et al., 2009).

MRSA isolates originating from animals have been shown to hold important genes of resistance which could be transferred to less pathogenic human strains, but well adapted, in a nasal co-colonization and resulting in new human lineages, for example (Springer et al., 2009). The gene czrC, which confers resistance to cadmium and zinc, was determined in isolates of MRSA CC398 of SCCmec type V originating from 23 (74%) pigs and 24 (48%) humans from Denmark. It is suggested that resistance to heavy metals can play a role in coselection of MRSA, because it was strongly related to the clone CC398 (Cavaco et al., 2010).

In lineages MRSA ST398 and MSSA ST9 isolated from pigs in Germany, the major reservoir of these lineages, the gene cfr was found, a gene of multi-resistance to the drugs phenicol/ lincosamine/ oxazolidinone/pleuromutilin/streptogramin A. The risk of its transference to humans with exposure to pig farms is of concern since these lineages can colonize and cause infections in humans (Kehrenberg et al., 2009).

Genes encoding virulence factors can also be carried by animal strains. An example is the PVL gene, encoding an important virulence factor related to MRSA, rarely reported in animals, but can be found in companion animals. PVL-positive CA-MRSA has been reported in cats, dogs, rabbits, birds, bats, turtles, pigs and cattle. The strains associated with pigs that have been rapidly disseminating are currently PVL-negative (Morgan, 2008).

As for animal origin food, meats of several animal species have been evaluated for detection of MRSA. The contamination has been reported in meats of turkey, chicken, veal, pork, beef and lamb. The majority of the isolates were non typeable MRSA. Considering the low number of non typeable MRSA in patients, the role of food products in disseminating MRSA seems to have been overlooked (Morgan, 2008).

5. MRSA DETECTION IN ANIMALS

Animal studies of MRSA are limited by certain research bottlenecks compared to human studies, including a lack of standardization regarding culture methodology, susceptibility testing, definition of genetic profile and sampling methods, which ultimately renders comparison difficult. A mass animal screening is logistically difficult, expensive, and impractical in many situations. Moreover, physical challenges exist: for example, carrying out a nasal swab in cats is only possible with appropriate animal restraints (Morgan, 2008).

Despite these drawbacks, several studies have been carried out, as more detailed studies of epidemiological aspects of animal MRSA are indispensable, primarily in food-producing animals of particular concern. These animals are not only reported as the primary source of a recently emergent new type of MRSA, LA-MRSA, but studies also suggest that they are involved in transmission of other strains of MRSA between animals and humans (Vanderhaeghen et al., 2010). MRSA screening must be performed in all diagnostic laboratories, even if done through disk diffusion with oxacillin, which is better than methicillin related to the resistance to degradation and the detection of heteroresistant strains (Weese et al., 2004).

MRSA detection in animals has been performed, generally, with isolation from samples from nasal and oral mucosae and perineum in small animals (Loeffler et al., 2005; Nienhoff et al., 2009), samples from milk in cows (Kwon et al., 2005), samples from nasal, oral and/or perineal swab, in pigs and horses (Baptiste et al., 2005; Voss et al., 2005; Huijsdens et al., 2006), samples from nose and cloacae in poultries (Nemati et al., 2008) and samples from meats (Lozano et al., 2009).

The samples were pre-enriched in different selective mediums with the aim of increasing the sensitivity of the culture technique. Lozano et al. (2009), to detect MRSA in meat, pre-enriched the samples in BHI broth (brain heart infusion) containing 6.5% of sodium chloride at 35°C for 24 h. An aliquot of each growth was seeded on ORSAB plates (oxacillin resistance screening agar base) with oxacillin (2mg/L) and incubated at 35°C for 36 h. Van Duijkeren et al. (2010) compared two methods of culture technique; the first was Müeller Hinton Broth containing 6.5% sodium chloride, which after overnight incubation at 37°C, was transferred to phenol red mannitol broth with 5µg/ml of ceftizoxime and 75µg/ml of aztreonam and after overnight incubation at 37°C was plated onto sheep blood agar and brilliance MRSA agar. In the second method, Tryptone Soy Broth was used containing 4% sodium chloride, 1% mannitol, 16µg/ml of phenol red, 50µg/ml of aztreonam and 5µg/ml ceftizoxime, plating onto sheep blood agar and MRSA brilliance agar after 48 hours of incubation at 37°C. The first method, with the preenrichment containing higher salt concentration, presented better results. Weese et al. (2011) used as pre-enrichment a broth containing tryptone, sodium chloride, mannitol, yeast extract and incubated at 35°C for 24 h to evaluate MRSA presence in feedlot cattle, as well the nose samples as feces. An aliquot of the growth was inoculated onto MRSA Chromogenic agar and incubated at 35°C for 48 hours.

Many methodologies have been employed to identify and characterize strains, ranging from phenotypic to genotypic. Both present advantages and disadvantages and must be performed in accordance with the need of the study and the material and personnel available in each laboratory. Relative speed and the reliability are the desirable characteristics in both methods, because the choice of treatment and infection control measures are determined by the results of such testing (Kaya et al., 2009). Phenotypic methods at first are more accessible and nearly always cheaper; however they depend on the characteristic expression and visualization that cannot occur or be reduced, as for example, by environmental influences and/or regulatory genes (Berger-Bächi, 2002; Mohanasoundaram & Lalitha, 2008).

Methicillin-resistant S. aureus can be identified through different genotypic methods, such as species-specific primers for detection of DNA fragment of S. aureus-specific (van Duijkeren et al., 2010, as cited in Martineau et al., 1998) and with gene mecA (Murakami et al., 1991) by PCR (polymerase chain reaction), or the detection of DNA fragment of S. aureus-specific and gene mecA by multiplex PCR (Huijsdens et al., 2006), for example.

To determine the MRSA clones involved, in the beginning of 1990s the pulsed-field gel electrophoresis (PFGE) of genomic SmaI macrorestriction fragments were introduced and still represents the gold standard with respect to discriminatory power. The clonal groups determined by cluster analysis through PFGE are largely congruent with those defined by MLST. However, with the presence of some lineages of special interest (for example ST398) that are non typeable by the standard restriction enzyme SmaI, other methods have been used (Cuny et al., 2010).

As observed in the majority of studies discussed in this chapter, from genotypic methods for classification of predominant strains and determination of evolutionary pathways, MLST and spa typing have been the most widely employed: the first because it is an unambiguous discriminatory method for studying MRSA epidemiology and evolution, with results that can be truly portable between laboratories (Enright, 2003) and the second method to indicate genetic microvariation permitting investigation of outbreaks or accomplishment of phylogenetic analysis (Koreen et al., 2004). MLST characterize bacteria isolates unambiguously using the sequences of internal fragments of seven "housekeeping" genes, being a discriminatory method which permits that related strains recovered from different countries be quickly identified (Enright et al., 2002). The spa type identified by DNA sequence analysis of the X region of the protein A gene (spa) is less expensive, timeconsuming, and error prone than multilocus techniques, such as MLST (Shopsin et al., 1999; Koreen et al., 2004).

The classification by MLST permits that the genomes of strains deposited in the GenBank database to be compared to establish their evolution and characteristic features. Comparing the genome of CA-MRSA and HA-MRSA from the same clonal lineage as well as their most probable MSSA ancestor, about 78% of the genes are conserved, and the remaining 22% comprise an "accessory genome" including genomic islands, pathogenicity islands, prophages, integrated plasmids, and transposons. However, comparing the S. aureus genome from cattle mastitis (ST151) with human S. aureus, it has been demonstrated that this bovine clone probably evolved from a common ancestor by acquiring foreign DNA. Subsequent microarray studies on recent epidemic strains of bovine origin (such as ST97) also revealed the presence of mobile genetic elements absent from S. aureus of human origin. Even with these impressive insights into the evolution of S. aureus, research has not yet provided many clues on the adaptation of the pathogen to the host, since only limited data are available (Cuny et al., 2010). In this respect, further research is needed to address these gaps, as well as to better understand the evolution of these strains in humans and animals.

6. MRSA PREVENTION AND CONTROL IN ANIMALS

All studies on animal MRSA have helped establish critical measures for its control. Numerous reports on MRSA control in humans have been published and many of the principles may also be applied to control in animals. However, caution is necessary for extrapolating these human guidelines to animals, as disease epidemiology can differ significantly (Leonard & Markey, 2008).

It has been observed that exposure to antimicrobials is a risk factor for the acquisition and dissemination of MRSA in humans and also most probably in animals. In this respect, strategies for prevention and management of MRSA in animals should be, as much as possible, related to the use of antimicrobials. If the antimicrobial treatment is necessary in individual cases for the sake of animal welfare, the risk of the emergence of wider resistance in MRSA strains colonizing animals needs to be managed, especially considering zoonotic aspects. Options to manage this risk include the non-use of antimicrobials except as a last resort strategy, decolonization in humans, isolation of animals during treatment, and monitoring the effects of treatment in strain resistance through selective culture and susceptibility tests (CATRY et al., 2010).

6.1 General preventive and control measures

Good hygiene is an important general preventive and control measure, both in homes and human and animal healthcare environments, because environmental contamination with MRSA acts as a reservoir for infection. Known MRSA-positive animals should be nursed apart from other animals, with strict washing of the hands, gloves and gowns if in close contact. Recording the history of contact with human or animal MRSA, as well as an early culture of a wound non-responsive to first-line therapy allows for earlier recognition of MRSA and its appropriate management. Furthermore, when faced with repeated and inexplicable failure of human decolonization, clinicians can investigate nearby exposure to animals and birds that could be the reservoirs (Morgan, 2008).

Below, are some precolonized specific measures cited from (Catry et al., 2010).

6.1.1 Specific measures for livestock animals

Reduction of antimicrobial selective pressure in livestock by avoiding routine mass medication

Prevention of transmission of MRSA between and within the farms with sanitary measures of control between herds and during transportation

Identification and isolation of animals to minimize the risk for zoonotic infection

Use of contact precautions such as protective outerwear, overalls, aprons or coats and boots or overshoes that are not worn elsewhere

Protective outerwear and all the items handled during the treatment of MRSA-positive animals should be considered potentially contaminated

Hands can be hygienically cleaned with alcohol gel pouches, which are essential but need to be used correctly

Proper cleaning and disinfection of contaminated environments, including transport vehicles. Special attention should be paid to dust in stables

Animal owners should be informed about the risks and necessary precautions.

6.1.1.1 Reducing carriers on MRSA-positive livestock farms

Control and/or treatment of colonized and infected animals with or without antimicrobials is necessary for the reduction of carriers.

Th1e affected animals need to be immediately separated from healthy animals. In extreme cases culling of infected animals is a further option.

Milk of animals with mastitis by MRSA must be destroyed, and in some cases the infected quarter must be prematurely dried-off.

If the antimicrobial treatment is chosen, it is necessary to evaluate its risk-benefit compared with other alternatives. The choice of antimicrobials should always be based on a susceptibility test, and all precautions should be taken that the drug reaches the infected site with appropriate concentrations.

6.1.2 Specific measures for horses

The options of control for colonized horses, as well as livestock animals, include the nonantimicrobial management and the antimicrobial treatment of colonized or infected horses. In a Canadian study, two farms with horses colonized by MRSA drastically reduced the number of affected animals with active screening and strict implementation of control protocols of infection, without the use of antimicrobial therapy. Antimicrobial treatment must be applied only if the colonization is persistent or in cases in which control measures are impossible.

Preventive measures for the infected animals are the same for the previously mentioned livestock animals. In equine hospitals, MRSA management for veterinary practices, guidelines stipulated by the British Small Animal Veterinary Association (BSAVA), for example, can also be applicable. In the confirmed or suspected cases of infection by MRSA, horses must be isolated and treated as if disseminating a nosocomial and zoonotic agent. It is necessary to take precautions with staff hygiene as well, using protective barriers, such as gloves, aprons, and boots. Moreover, the entire animal must be evaluated before being admitted to the hospital to ensure prevention of MRSA dissemination.

6.1.3 Specific measures for companion animals

In colonized animals, it not recommended to decolonize animals having mucosae colonization with MRSA. And it is observed that apparently some pets eliminate the carriage of MRSA spontaneously if the re-colonization is prevented. Antimicrobial therapy is only indicated in exceptionally persistent cases or when control measures are impossible., If the infection is local in infected animals, meticulous management of the wound can avoid the necessity of antimicrobial use. The risk of resistance development, the susceptibility profile of the isolate, the severity of the infection and the presence of systemic disease (fever, leukocytosis), and the patient's underlying disease or any comorbidity must be taken into account when choosing the antimicrobial treatment. In deciding to treat or eventually euthanize, veterinarians must consider the national veterinary guidelines available.

Preventive measures include the strict control practices of the infection in homes, particularly frequent hand washing hygiene, avoiding high-risk contact to minimize the chance of becoming colonized and isolating the animal and other pets. Gloves, masks and eye protection must be used to attend patients, as well as planning surgeries to avoid risk for infection, removing permanent urinary catheters as soon as possible and preventing complications from intravenosous and urinary catheters with appropriate asepsis.

7. CONCLUSION

This paper summarizes a wide range of information and findings from the literature on MRSA in animals and humans in contact with them. Notwithstanding, there is enormous potential for new research aiming to conclusively address certain unknown questions such as, at which point does an infection play the role of a zoonosis or humanosis? This question has yet to be answered because several animal species are involved, with distinct characteristics of transmission and isolated clones, raising appropriate concern for MRSA in animals.

Steadfast vigilance of MRSA in samples of animal origin in laboratory diagnoses is essential for: consistent and thorough monitoring of the evolution and dissemination of these strains; elucidating characteristics that determine a predilection for a determined host; determing transmission routes; identifying resistance and virulence genes received by these new lineages; and distinguishing molecular markers that allow for discriminating between CAMRSA, HA-MRSA and LA-MRSA.

Appropriate and effective measures of control and prevention must be better determined and applied to each situation and country, according to previously reported guidelines, aiming to minimize risks to humans, since these strains have housed new virulence and resistance genes which can be transferred to human strains. Veterinarians play an important role in public health, in controlling this pathogen through measures appropriately applied in veterinary medicine, namely, the rational use of antimicrobials and appropriate management of infected animals, together with other health professionals, for prevention of MRSA dissemination.

ACKNOWLEDGEMENT

The support by FAPESP is gratefully acknowledged.

REFERENCES

1. Baptiste, K.E.; Williams, K.; Williams, N.J.; Wattret, A.; Clegg, P.D.; Dawson, S.; Corkill, J.E.; O'Neill, T. & Hart, C.A. (2005). Methicillin-resistant staphylococci in companion animals. Emerging Infectious Diseases, Vol.11, No.12, pp.1942-1944, ISSN 1080-6059

2. Bens, C.C.P.M.; Voss, A. & Klaassen, C.H.W. (2006). Presence of a novel DNA methylation enzyme in methicillin-resistant Staphylococcus aureus isolates associated with pig farming leads to uninterpretable results in standard pulsed-field gel electrophoresis analysis. Journal of Clinical Microbiology, Vol.44, No.5, pp.1875-1876, ISSN 1098-660X

3. Berger-Bächi, B. (2002). Resistance mechanisms of gram-positive bacteria. International Journal of Medical Microbiology, Vol.292, No.1, pp.27-35, ISSN 1438-4221

4. Catry, B.; van Duijkeren, E.; Pomba, M.C.; Greko, C.; Moreno, M.A.; Pyörala, S.; Ruzauskas, M.; Sanders, P.; Threlfall, E.J.; Ungemach, F.; Törneke, K.; Muňoz-Madero, C. & Torren-Edo, J. (2010). Reflection paper on MRSA in food-producing and companion animals: epidemiology and control options for human and animal health. Epidemiology and Infection, Vol.138, No.5, pp.626-644, ISSN 0950-2688

5. Cavaco, L.M.; Hasman, H.; Stegger, M.; Andersen, P.S.; Skov, R.; Fluit, A.C.; Ito, T. & Aarestrup, F.M. (2010). Cloning and occurrence of crzC, a gene conferring cadmium and zinc resistance in methicillin-resistant Staphylococcus aureus CC398 isolates.

Antimicrobial Agents and Chemotherapy, Vol.54, No.9, pp.3605-3608, ISSN 0066-4804

6. Cuny, C.; Friedrich, A.; Kozytska, S.; Layer, F.; Nübel, U.; Ohlsen, K.; Strommenger, B.; Walther, B.; Wieler, L. & Witte, W. (2010). Emergence of methicillin-resistant Staphylococcus aureus (MRSA) in different animal species. International Journal of Medical Microbiology, Vol.300, No.2-3, pp.109-117, ISSN 1438-4221

7. David, M.Z. & Daum, R.S. (2010). Community-associated methicillin-resistant Staphylococcus aureus: epidemiology and clinical consequences of an emerging epidemic. Clinical Microbiology Reviews, Vol.23, No.3, pp.616-687, ISSN 0893-8512

8. Enright, M.C.; Robinson, D.A.; Randle, G.; Feil, E.J.; Grundmann, H. & Spratt, B.G. (2002). The evolutionary history of methicillin-resistant Staphylococcus aureus. PNAS, Vol.99, No.11, pp.7687-7692, ISSN 1091-6490

9. Enright, M.C. (2003). The evolution of a resistant pathogen – the case of MRSA. Current Opinion in Pharmacology, Vol.3, No.5, pp.474-479, ISSN 1471-4892

10. Feβler, A.; Scott, C.; Kadlec, K.; Ehricht, R.; Monecke, S. & Schwarz, S. (2010). Characterization of methicillin-resistant Staphylococcus aureus ST398 from cases of bovine mastitis. Journal of Antimicrobial Chemotherapy, Vol.65, No.4, pp.619-625, ISSN 0305-7453

11. Graveland, H.; Wagenaar, J.A.; Heesterbeek, H.; Mevius, D.; Van Duijkeren, E. & Heederik, D. (2010). Methicillin resistant Staphylococcus aureus ST398 in veal calf farming: human MRSA carriage related with animal antimicrobial usage and farm hygiene. Plos One, Vol.5, No.6, e10990, pp.1-5, ISSN 1932-6203

12. Guinane, C.M.; Zakour, N.L.B.; Tormo-Mas, M.A.; Weinert, L.A.; Lowder, B.V.; Cartwright, R.A.; Smyth, D.S.; Smyth, C.J.; Lindsay, J.A.; Gould, K.A.; Witney, A.; Hinds, J.; Boolback, J.P.; Rambaut, A.; Penadés, J.R. & Fitzgerald, J.R. (2010). Evolutionary genomics of Staphylococcus aureus reveals insights into the origin and molecular basis of ruminant host adaptation. Genome Biology and Evolution, v.2, pp.454-466, ISSN 1759-6653

13. Horgan, M.; Abbott, Y.; Lawlor, P.G.; Rossney, A.; Coffey, A.; Fitzgerald, G.F.; Mcauliffe, O. & Ross, R.P. (2010). A study of the prevalence of methicillin-resistant Staphylococcus aureus in pigs and in personnel involved in the pig industry in Ireland. The Veterinary Journal, in press, ISSN 1090-0233

14. Huijsdens, X.W.; van Dijke, B.J.; Spalburg, E.; van Santen-Verheuvel, M.G.; Heck, M.E.O.C.; Pluister, G.N.; Voss, A.; Wannet, W.J.B. & de Neeling, A.J. (2006). Communityacquired MRSA and pig-farming. Annals of Clinical Microbiology and Antimicrobials, Vol.5, No.26, ISSN 1476-0711

15. Hunter, P.A.; Dawson, S.; French, G.L.; Goossens, H.; Hawkey, P.M.; Kuijper, E.J.; Nathwani, D.; Taylor, D.J.; Teale, C.J.; Warren, R.E.; Wilcox, M.H.; Woodford, N.; Wulf, M.W. & Piddock, L.J.V. (2010). Antimicrobial-resistant pathogens in animals and man: prescribing, practices and policies. Journal of Antimicrobial Chemotherapy, Vol.65, suppl.1, pp.i3-i17, ISSN 0305-7453

16. Juhász-Kaszanyitzky, E.; Jánosi, S.; Somogyi, P.; Dán, A.; Bloois, L.G.; van Duijkeren, E. & Wagenaar, J.A. (2007). MRSA transmission between cows and humans. Emerging Infectious Diseases, Vol.13, No.4, pp.630-632, ISSN 1080-6059

17. Kaya, E.G.; Karakoç, E.; Yağci, S. & Yücel, M. (2009). Evaluation of phenotypic and genotypic methods for detection of methicillin resistance in Staphylococcus aureus. African Journal of Microbiology Research, Vol.3, No.12, pp.925-929, ISSN 1996-0808

18. Kehrenberg, C.; Cuny, C.; Strommenger, B.; Schwartz, S. & Witte, W. (2009). Methicillinresistant and –susceptible Staphylococcus aureus strains of clonal lineages ST398 and ST9 from swine carry the multidrug resistance gene cfr. Antimicrobial Agents and Chemotherapy, Vol.53, No.2, pp.779-781, ISSN 0066-4804

19. Kitai, S.; Shimizu, A.; Kawano, J.; Sato, E.; Nakano, C.; Uji & T. Kitagawa, H. (2005). Characterization of methicillin-resistant Staphylococcus aureus isolated from retail raw chicken meat in Japan. The Journal of Veterinary Medical Science, Vol.67, No.1, pp.107-110, ISSN 1347-7439

20. Koreen, L.; Ramaswamy, S.V.; Graviss, E.A.; Naidich, S.; Musser, J.M. & Kreiswirth, B.N. (2004). spa Typing method for discriminating among Staphylococcus aureus isolates: implications

for use of a single marker to detect genetic micro- and macrovariation. Journal of Clinical Microbiology, Vol.42, No.2, pp.792-799, ISSN 1098-660X

21. Kwon, H.H.; Park, K.T.; Moon, J.S.; Jung, W.K.; Kim, S.H.; Kim, J.M.; Hong, S.K.; Koo, H.C.; Joo, Y.S. & Park, Y.H. (2005). Staphylococcal cassette chromosome mec (SCCmec) characterization and molecular analysis for methicillin-resistant Staphylococcus aureus and novel SCCmec subtype IVg isolated from bovine milk in Korea. Journal of Antimicrobial Chemotherapy, Vol.56, No.4, pp.624-632, ISSN 0305-7453

22. Lee, J.H. (2003). Methicillin (Oxacillin)-resistant Staphylococcus aureus strains isolated from major food animals and their potential transmission to humans. Applied and Environmental Microbiology, Vol.69, No.11, pp.6489-6494, ISSN 0099-2240

23. Leonard, F.C. & Markey, B.K. (2008). Methicillin-resistant Staphylococcus aureus in animals: a review. The Veterinary Journal, Vol.175, No.1, pp.27036, ISSN 1090-0233

24. Li, S.; Skov, R.L.; Han, X.; Larsen, A.R.; Larsen, J.; Sørum, M.; Wulf, M.; Voss, A.; Hiramatsu, K. & Ito, T. (2011). Novel types of staphylococcal cassette chromosome mec elements identified in CC398 methicillin resistant Staphylococcus aureus strains. Antimicrobial Agents and Chemotherapy, Vol.55, No.2, pp.3046-3050, ISSN 0066-4804

25. Loeffler, A.; Boag, A.K.; Sung, J.; Lindsay, J.A.; Guardabassi, L.; Dalsgaard, A.; Smith, H.; Stevens, K.B. & Lloyd, D.H. (2005). Prevalence of methicillin-resistant Staphylococcus aureus among staff and pets in a small animal referral hospital in the UK. Journal of Antimicrobial Chemotherapy, Vol.56, No.4, pp.692-697, ISSN 0305-7453

26. Loeffler, A.; Kearns, A.M.; Ellington, M.J.; Smith, L.J.; Unt, V.E.; Lindsay, J.A.; Pfeiffer, D.U. & Lloyd, D.H. (2009). First isolation of MRSA ST398 from UK animals: a new challenge for infection control teams? Journal of Hospital Infection, Vol.72, No.3, pp.269-271, ISSN 0195-6701

27. Loeffler, A. & Lloyd, D.H. (2010). Companion animals: a reservoir for methicillin-resistant Staphylococcus aureus in the community?

Epidemiology and Infection, Vol.138, No.5, pp.595-605, ISSN 0950-2688

28. Lozano, C.; López, M.; Gómez-Sanz, E.; Ruiz-Larrea, F.; Torres, C. & Zarazaga, M. (2009). Detection of methicillin-resistant Staphylococcus aureus ST398 in food samples of animal origin in Spain. Journal of Antimicrobial Chemotherapy, Vol.64, No.6, pp.1325- 1346, ISSN 0305-7453

29. Lozano, C.; Aspiroz, C.; Ezpeleta, A.I.; Gómez-Sanz, E.; Zarazaga, M. & Torres, C. E. (2011). Empyema caused by MRSA ST398 with atypical resistance profile, Spain. Emerging Infectious Diseases, Vol.17, No.1, ISSN 1080-6059

30. Mohanasoundaram, K.M. & Lalitha, M.K. (2008). Comparison of phenotypic versus genotypic methods in the detection of methicillin resistance in Staphylococcus aureus. Indian Journal of Medical Research, Vol.127, No.1, pp.78-84, ISSN 0971-5916

31. Morgan, M. (2008). Methicillin-resistant Staphylococcus aureus and animals: zoonosis or humanosis? Journal of Antimicrobial Chemotherapy, Vol.62, No.6, pp.1181-1187, ISSN 0305-7453

32. Murakami, K.; Minamide, W.; Wada, K.; Nakamura, E.; Teraoka, H. & Watanabe, S. (1991). Identification of methicillin-resistant strains of staphylococci by polymerase chain reaction. Journal of Clinical Microbiology, Vol.29, No.10, pp.2240-2244, ISSN 1098- 660X

33. Nienhoff, U.; Kadlec, K.; Chaberny, I.F.; Verspohl, J.; Gerlach, G.F.; Schwarz, S.; Simon, D. & Nolte, I. (2009). Transmission of methicillin-resistant Staphylococcus aureus strains between humans and dogs: two case reports. Journal of Antimicrobial Chemotherapy, Vol.64, No.3, pp.660-662, ISSN 0305-7453

34. Persoons, D.; van Hoorebeke, S.; Hermans, K.; Butaye, P.; Kruif, A.; Haesebrouck, F. & Dewulf, J. (2009). Methicillin-resistant Staphylococcus aureus in poultry. Emerging Infectious Diseases, Vol.15, No.3, pp.452-453, ISSN 1080-6059

35. Springer, B.; Orendi, U.; Much, P.; Höger, G.; Ruppitsch, W.; Krziwanek, K.; Metz-Gercek, S. & Mittermayer, H. (2009). Methicillin-resistant Staphylococcus aureus: a new zoonotic agent? Wiener klinische Wochenschrift, Vol.121, No.3-4, pp.86–90, ISSN 0043- 5325

36. Türkyılmaz, S.; Tekbıyık, S.; Oryasin, E. & Bozdogan, B. (2010). Molecular epidemiology and antimicrobial resistance mechanisms of methicillin-resistant Staphylococcus aureus isolated from bovine milk. Zoonoses and Public Health, Vol.57, No.3, pp.197-203, ISSN 1863-1959

37. Van Duijkeren, E.; Box, A.T.A.; Heck, M.E.O.C.; Wannet, W.J.B. & Fluit, A.C. (2004). Methicillin-resistant staphylococci isolated from animals. Veterinary Microbiology, Vol.103, No.1-2, pp.91-97, ISSN 0378-1135

38. Van Loo, I.; Huijsdens, X.; Tiemersma, E.; De Neeling, A.; van de Sande-Bruinsma, N.; Beaujean, D.; Voss, A. & Kluytmans, J. (2007). Emergence of methicillin-resistant Staphylococcus aureus of animal origin in humans. Emerging Infectious Diseases, Vol.13, No.12, pp.1834-1839, ISSN 1080-6059

39. Vanderhaeghen, W.; Hermans, K.; Haesebrouck & F., Butaye, P. (2010). Methicillin-resistant Staphylococcus aureus (MRSA) in food production animals. Epidemiology and Infection, Vol.138, No.5, pp.606-625, ISSN 0950-2688

40. Voss, A.; Loeffen, F.; Bakker, J.; Klaassen, C. & Wulf, M. (2005). Methicillin-resistant Staphylococcus aureus in pig farming. Emerging Infectious Diseases, Vol.11, No.12, pp.1965-1966, ISSN 1080-6059

41. Wagenaar, J.A.; Yue, H.; Pritchard, J.; Broekhuizen-Stins, M.; Huijsdens, X.; Mevius, D.J.; Bosch, T. & van Duijkeren, E. (2009).Unexpected sequence types in livestock associated methicillin-resistant Staphylococcus aureus (MRSA): MRSA ST9 and a single locus variant of ST9 in pig farming in China. Veterinary Microbiology, Vol.139, No.3-4, pp.405-409, ISSN 0378-1135

42. Weese, J.S. (2004). Methicillin-resistant Staphylococcus aureus in horses and horse personnel. The Veterinary Clinics of North America. Equine Practice, Vol.20, No.3, pp.601-613, ISSN 0749-0739

43. Weese, J.S.; Avery, B.P. & Reid-Smith, R. J. (2010). Detection and quantification of methicillin-resistant Staphylococcus aureus (MRSA) clones in retail meat products. Letters in Applied Microbiology, Vol.51, No.3 ,pp.338–342, ISSN 0266-8254

44. Weese, J.S.; Hannon, S.J.; Booker, C.W.; Gow, S.; Avery, B.P. & Reid-Smith, R.J. (2011). The prevalence of methicillin-resistant Staphylococcus aureus colonization in feedlot cattle. Zoonoses and Public Health, in press, ISSN 1863-1959

45. Wulf, M.W.H.; Sørum, M.; van Nes, A.; Skov, R.; Melchers, W.J.G.; Klaassen, C.H.W. & Voss, A. (2008). Prevalence of methicillin-resistant Staphylococcus aureus among veterinarians: an international study. Clinical Microbiology and Infection, Vol.14, No.1, pp.29-34, ISSN 1198-743X

CHAPTER 10

A new bovine conjunctiva model shows that Listeria monocytogenes invasion is associated with lysozyme resistance

Jessica Warren a , A. Rhys Owen a , Amy Glanvill a , Asher Francis b , Grazieli Maboni a , Rodrigo J. Nova a , Wendela Wapenaar a , Catherine Rees b , Sabine To¨temeyer a, *

a School of Veterinary Medicine and Science, University of Nottingham, Sutton Bonington Campus, Loughborough LE12 5RD, UK

b School of Biosciences, University of Nottingham, Sutton Bonington Campus, Loughborough LE12 5RD, UK

ABSTRACT

Listerial keratoconjunctivitis ('silage eye') is a wide spread problem in ruminants causing economic losses to farmers and impacts negatively on animal welfare. It results from direct entry of *Listeria monocytogenes* into the eye, often following consumption of contaminated silage. An isolation protocol for bovine conjunctival swabbing was developed and used to sample both infected and healthy eyes bovine eyes (n = 46). *L. monocytogenes* was only isolated from one healthy eye sample, and suggests that this organism can be present without causing disease. To initiate a study of this disease, an infection model was developed using isolated conjunctiva explants obtained from cattle eyes post slaughter. Conjunctiva were cultured and infected for 20 h with a range of *L. monocytogenes* isolates (n = 11), including the healthy bovine eye isolate and also strains isolated from other bovine sources, such as milk or clinical infections. Two *L. monocytogenes* isolates (one from a healthy eye and one from a cattle abortion) were markedly less able to invade conjunctiva explants, but one of those was able to efficiently infect Caco2 cells indicating that it was fully virulent. These two isolates were also significantly more sensitive to lysozyme compared to most other isolates tested, suggesting that lysozyme resistance is an important factor when infecting bovine conjunctiva. In conclusion, we present the first bovine conjunctiva explant model for infection studies and demonstrate that clinical *L. monocytogenes* isolates from cases of bovine keratoconjunctivitis are able to infect these tissues.

KEYWORDS

Listeria monocytogenes; Silage eye; Keratoconjunctivitis; *Ex vivo* organ culture;Lysozyme resistance

1. INTRODUCTION

Listeria monocytogenes is an intracellular pathogen which has been extensively studied as a cause of human food borne infections associated with a high mortality rate (McLauchlin et al., 2014). In both man and animals *L. monocytogenes* is able to infect a wide range of tissues including the spleen, liver and brain and can cross the placenta to cause abortions (Cossart and Toledo-Arana, 2008). It is found widely in the farm environment and its ability to use plant sugars as part of its saprophytic lifestyle means that it can reach levels as high as 10^8 cfu g^{-1} wet weight in some poorly fermented silages when low pH does not suppress growth of the organism (Wiedmann et al., 1996). Listerial keratoconjunctivitis and uveitis ('silage eye') are common problems in ruminants in the UK, which negatively impact on animal welfare and causes economic losses to farmers (Erdogan, 2010). A strong correlation has been established between silage eye and the use of big bale silage and silage feeding in ring feeders and these infections are believed to occur when the organism directly enters the eye, possibly facilitated by corneal abrasions (Erdogan, 2010 and Revathi et al., 2011). Treatment of listerial keratoconjunctivitis includes parenteral and/or topical use of antibiotics generally resulting in recovery within two weeks, however, antibiotic resistance has been noted as an increasing problem (Erdogan, 2010).

There have been increasing field reports of silage eye in ruminants since the late 1980s (Erdogan et al., 2001) but very little research to try and understand the route of transmission, carrier state in the eye/conjunctiva or pathology of the disease. Currently the lack of understanding of silage eye infection routes and epidemiology reduces the development of treatment and preventative measures (Erdogan, 2010). One of the main defence mechanisms in bovine tears is lysozyme with concentrations in bovine tears of up to 580 μg ml^{-1} reported in the literature (Gionfriddo et al., 2000). Similar levels have been reported in sheep's eyes (600 μg ml^{-1}; Gionfriddo et al., 2000) but can be much higher in humans, and is reported to be up to 2.0 mg ml^{-1} (Sherman et al., 1994). Like many bacterial pathogens, *L. monocytogenes* is known to be naturally resistant to lysozyme by modifying its peptidoglycan (PG) structure so that it cannot be degraded by the enzyme. Three PG modifying enzymes involved in lysozyme resistance, peptidoglycan deacetylase PgdA, putative carboxypeptidase PbpX and acetyltransferase OatA, have been identified (Aubry et al., 2011, Boneca et al., 2007 and Burke et al., 2014) and also two regulators of gene expression, DegU (response regulator) and Rli31 (long non-coding RNA) which are believed to up-regulate the expression of these genes (Burke et al., 2014). The PG modifications carried out by PgdA and OatA also result in reduced host immune response by suppressing NOD1-dependent and toll-like receptor 2 (TLR2) IL-6 and interferon-β secretion as well as IL1β and IL12 mRNA expression (Aubry et al., 2011, Aubry et al., 2012 and Rae et al., 2011). Hence the high level of lysozyme resistance seen in many clinical isolates may be linked to a selective advantage conferred by an ability survive host lysozyme challenge and evade the host immune response.

The aim of this study was to develop a conjunctiva explant infection model to determine whether *L. monocytogenes* isolated from bovine keratoconjunctivitis cases have properties that allow better infection of conjunctiva in comparison to isolates from healthy cattle eyes, milk or other bovine clinical conditions.

Table 1. *Listeria monocytogenes* isolates used in this study.

Strain number	Source[a]	PCR serotype[b]	PCR lineage[c]	Source/reference
10403S	Skin Lesion	1/2a	II	Bishop and Hinrichs (1987)
AR008	Healthy eye	1/2a, 3a	II	This study
C00938	Silage eye	1/2a, 3a	II	APHA
R06262	Silage eye	1/2b, 3b	I	APHA
C02118	Silage eye	4b	I	APHA
LM7644	Abortion	1/2a, 3a	II	APHA
C08389	Abortion	1/2a, 3a	II	APHA
G03652	Meningitis	1/2b, 3b	I	APHA
LM4	Milk	1/2b, 3b	I	Lawrence et al. (1995)
LM6	Milk	4b	I	Lawrence et al. (1995)
LM7	Milk	4b	I	Lawrence et al. (1995)

APHA, Animal and Plant Health Agency.

[a] All isolates are from bovine sources except for the human isolate 10403S.

[b] Serotypes were determined using the PCR-based method of Doumith et al. (2004). This method in conjunction with the lineage typing cannot distinguish between serotypes 1/2a and 3a or 1/2b and 3b. However, serotypes 3a and 3b are not commonly isolated.

[c] Lineages were determined using the PCR-based method of Ward et al. (2004).

2. MATERIALS AND METHODS

2.1. Bacterial culture

L. monocytogenes strains used in this study are listed in Table 1. Bacteria were cultured overnight (approximately 17 h) at 37 °C in Heart infusion (HI) broth (Oxoid) or on HI agar plates. Optical density (600 nm) was used as an estimate of cell number and then cfu ml^{-1}ascertained by viable count. Haemolytic activity was assessed by growth on sheep blood agar plates (Oxoid) after incubation at 37 °C overnight.

2.2. Sample collection

Clinical samples were collected by veterinary practitioners and ethical approval was obtained from the ethics committee of the School of Veterinary Medicine and Science, University of Nottingham. The swabs (Culture Swab Transport System, VWR International) were placed into the conjunctival sac and moved in a lateral and medial direction. Eye swabs were also taken from cattle heads which had been decapitated and facial muscles and skin removed in the slaughter process at abattoir or after transport to School of Veterinary Medicine and Science (SVMS). In those cases, the eye surface was also sampled using the same swab. Only bovine eyes that had a minimum area of skin of approximately 3 cm around the eye left were swabbed to reduce the microbial contamination through damage of the eyelids or conjunctiva during carcass processing. All swabs were stored at 4 °C overnight before culturing.

2.3. *Listeria monocytogenes* isolation from eye swabs

For direct plating, swabs were streaked onto Oxford agar (Oxoid) and plates incubated at 37 °C for 24–48 h. Any black/brown colonies were purified by streaking onto Brilliance Listeria agar (Oxoid) and incubated at 37 °C for 24–48 h. To detect low number of cells and adaptation of the ISO 112090-1 food sampling protocol was used. The same swabs used for direct plating were placed in 20 ml of Fraser broth (Oxoid) and vortexed for 30 s. After 1 h static incubation at 37 °C, the sample was divided into two portions and incubated at either 37 °C for 24–48 h or 4 °C for 10 d. Samples (100 µl) from any Fraser broth cultures displaying a colour change to brown/black after incubation were plated onto Brilliance Listeria agar. Gram stain was performed on all isolates giving the characteristic blue/green colonies surrounded by a zone of clearing. All Gram-positive rod shaped bacteria were further grown on to 5% sheep blood agar plates to identify weak β-haemolysis.

2.4. Multiplex PCR assay for *Listeria monocytogenes* serotyping

Multiplex PCR was performed in order to separate the four major serovars (1/2a, 1/2b, 1/2c, and 4b) and three main lineages (I, II, III) of *L. monocytogenes* (Doumith et al., 2004 and Ward et al., 2004). To prepare template DNA, three to six colonies were picked from Heart Infusion (HI) agar plates and mixed into 1 ml of sterile water, incubated at 90 °C for 10 min and then chilled on ice for 10 min and 1 μl of this was added to each PCR reaction.

2.5. Bovine conjunctiva explant culture

Conjunctiva samples were taken from cattle heads obtained from a commercial abattoir. Conjunctiva were dissected by cutting in the medial canthus area towards the orbit and from there a cut was made from cranial to caudal. The conjunctiva was then cut along the lateral canthus to meet with the previous cut. The palpebral tissue was then removed leaving a rectangular tissue section. Tissue sections were placed in phosphate buffered saline (PBS, Oxoid) and fat tissues were removed prior to using an 8 mm punch biopsy tool (Valu Biopsy Punch 8 mm 135963; National Veterinary Services, Stoke-on-Trent) to cut tissue into discs. Tissue disks were placed into basic medium (Dulbecco's Modified Eagle's Medium/Harris Nutrient Medium F12 (1:1 DME/F12 Modified, Sigma Aldrich) containing 5 μg ml^{-1} each of penicillin and streptomycin (P/S, Gibco Invitrogen, Paisley, England) and 0.63 μg ml^{-1} fungisome (Lonza). Before inoculation, tissue disks were placed in individual wells of a 24 well plate and washed with pre-warmed basic medium (37 °C) three times to remove any P/S and fungisome. Basic medium supplemented with 10% (v/v) foetal bovine serum and 2 mM L-glutamine (Gibco Invitrogen, England) was added to each well so that samples were just covered in medium.

2.6. Explant infection

L. monocytogenes cultures were grown overnight in HI broth 37 °C and inoculum of 10^9 cfu ml^{-1} was prepared. A sample (10 µl) of the inoculum was placed onto the epithelial surface of each tissue disc (MOI approx. 100; Infection density of 10^7 *Listeria* cells per 50 mm^2 and each explant was estimated to contain 3×10^5 conjunctival epithelial cells, based on an average cell size of 160 µm^2; Doughty and Hagan, 2013). Tissue discs were incubated in 5% CO_2 at 37 °C for 20 h, then the medium was replaced with basic medium containing gentamycin (100 µg ml^{-1}) to inactivate any remaining bacteria that had not entered the tissue. After 1 h samples were washed three times with PBS before being homogenised in PBS using stainless steel beads (5 mm, Qiagen Gmbh, Germany) and Bead Mill MM 301 (Retsch Gmbh, Germany) for 4 min at 30 movements s^{-1}. Bacteria were enumerated by plating onto HI agar plates.

2.7. Caco2 infections

The Caco2 infection assay was performed as described by Gaillard et al. (1987) with the following modification: Caco2 cells were infected with a multiplicity of infection (MOI) of 25 and 200 for 1 h, followed by incubation with gentamycin containing medium (100 µg ml^{-1}) for 1 h. Cells were washed with pre-warmed PBS and lysed with ice cold 0.5%, v/v Triton for 20 min prior to serial dilutions and plating on HI agar.

2.8. Lysozyme sensitivity assays

Bacteria were grown overnight, then resuspended in DME/F12 Modified medium and diluted to 10^5 cells ml^{-1} in 1:1 DME/F12 Modified medium supplemented with different concentrations of chicken egg Lysozyme (Sigma Aldrich; 1 µg ml^{-1} to 1 mg ml^{-1}). Samples were incubated for 24 h at 37 °C in a static incubator. Minimal inhibitory concentration (MIC) for lysozyme was determined by the dilution at which no visible growth was apparent. To identify the minimal bactericidal concentration (MBC), samples from wells where there was no visible growth were plated onto HI agar to determine if any viable bacterial cells remained.

For the lysozyme survival assay, a fixed concentration of lysozyme was used (580 µg µl^{-1}). The bacterial cells were prepared as before, but incubated in the presence or absence of lysozyme for 24 h in a 5% CO_2 incubator at 37 °C before the number of surviving bacteria was determined by viable count.

2.9. Statistical analysis

One way ANOVA (95% confidence interval) followed by Tukey's multiple comparison test was performed using GraphPad Prism6 software package.

3. RESULTS

3.1. Development of swabbing and isolation protocol from bovine conjunctiva

The limit of detection of the swabbing and isolation protocol was validated by using eyes in bovine heads collected post-slaughter. Conjunctiva were inoculated with a 30 µl inoculum containing between 1 and 10^6*L. monocytogenes* 10403S. The eyes were immediately swabbed as described previously, and then samples directly plated on to Oxford agar or inoculated into Fraser broth for enrichment and detection. The detection limit of the swabbing method was found to be 10^3 *L. monocytogenes* cells per conjunctiva.

3.2. Isolation of *L. monocytogenes* from bovine eyes without clinical signs of silage eye/keratoconjunctivitis

From 13 swabs from cattle with 'weepy' eyes obtained by veterinary practitioners, no*Listeria* were isolated. Based on colony morphology on Oxford and Listeria Brilliance agars, Gram stain reaction and weak haemolysis on sheep blood agar, two out of 33 swabs taken from clinically healthy eyes post-slaughter at an abattoir were identified as *L. monocytogenes*. However only one of these (AR008) survived freezing and could therefore be used for further experiments. This isolate, and all the clinical isolates obtained from the Animal and Plant Health Agency (APHA) culture collection, were serotyped using a multiplex PCR method (Doumith et al., 2004; Table 1). The strains used included representatives of serotypes 1/2a, 1/2b and 4b which cause 90% of human infections and are also commonly associated with animal infections (Bundrant et al., 2011).

3.3. Infectivity in conjunctiva explant model

Conjunctiva explants were infected with 10 different *L. monocytogenes* isolates, including the well characterised strain 10403S which was used as a reference strain. As the conjunctiva were taken from abattoir samples, a negative control set was also included to demonstrate that the eyes were not naturally infected with *L. monocytogenes* and no*Listeria* was recovered from these samples (Fig. 1). The other strains tested included the healthy eye isolate AR008, three isolates from cases of bovine conjunctivitis, one cattle abortion strain and three isolates from bovine milk (Table 1). The explants were inoculated with approximately 10^7 bacteria per sample. After incubation with the conjunctiva explants for 20 h, all *L. monocytogenes* isolates used were found to be able to invade the tissue, as indicated by the fact they were protected against a 1 h treatment with gentamicin (Kuhbacher et al., 2014). The level of bacteria recovered were 3–5 $\log_{10}$ cfu per explant (Fig. 1) but significant differences could be seen in the ability of the different *L. monocytogenes* to infect the explants. In particular two isolates, AR008 (healthy bovine eye) and LM7644 (bovine abortion) produced significantly lower levels of recovered intracellular bacteria compared to the other strains tested (Fig. 1).

3.4. Further investigation of strain characteristics

Haemolysin is a key enzyme required for the efficient release of internalised *L. monocytogenes* cells from the vacuole (Portnoy et al., 1992). To determine if variation in the ability of strains to infect the conjunctiva was associated with the level of production of haemolysin, all strains were tested by plating on sheep blood agar and were confirmed to be haemolytic, with all strains showing the same level of weak haemolysis as the control strain 10403S, none of the strains were either ahaemolytic or hyper-haemolytic (Table 2).

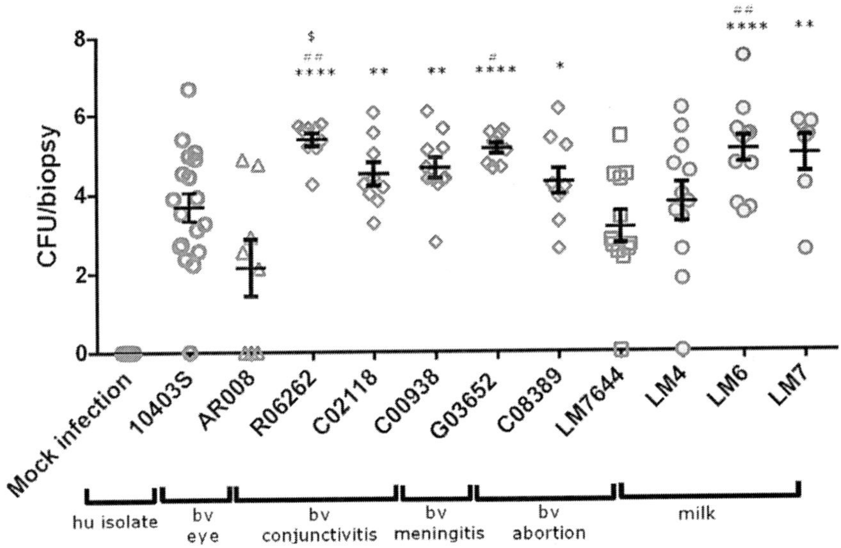

Figure 1. Infection of conjunctiva explants. Conjunctiva explants were inoculated with 10^7 bacteria per disc ($50\ mm^2$) and incubated for 20 h at 37 °C. For each strain a minimum of 7 independent experiments were performed; Mock infection $n = 18$, 10403S $n = 18$, LM4 $n = 12$, LM6 $n = 12$, LM7 $n = 7$, LM23 $n = 12$, LM7644 $n = 12$, AR008 $n = 8$, R06262 $n = 9$, C02118 $n = 9$, C00938 $n = 11$, G03652 $n = 9$, C08389 $n = 10$. All data points and mean ± SEM are shown. Explants infected with any of the *L. monocytogenes* strains showed significant differences to mock infection with $p < 0.0001$, except for AR008 with $p < 0.01$. $^{*}p < 0.05$; $^{**}p < 0.01$; $^{****}p < 0.0001$ compared to AR008; $^{\#}p < 0.05$; $^{\#\#}p < 0.01$ compared to LM7644;$^{\$}p < 0.05$ compared to 10403S (One way ANOVA & Tukey's multiple comparison test).

To determine if these strains were generally less able to infect human cells, the well-established *L. monocytogenes* Caco2 infection model was also used. Using this model AR008 showed the same ability to infect Caco2 cells as the control strain 10403S (Fig. 2). In contrast the abortion strain LM7644 and the milk isolates LM4 were both impaired in their ability to infect this cell type, with significantly lower number of bacteria recovered irrespective of the initial level of bacterial challenge (MOI = 25 or 200; Fig. 2).

Table 2. Haemolytic activity and lysozyme sensitivity.

Strain number	Source	Haemolysin production[a]	MIC[b] $\mu g\,ml^{-1}$ lysozyme	MBC[c] $\mu g\,ml^{-1}$ lysozyme
10403S	Skin Lesion	+	2500	3750
AR008	Healthy eye	+	78	2500
C00938	Silage eye	+	2500	2500
R06262	Silage eye	+	1250	5000
C02118	Silage eye	+	2500	2500
LM7644	Abortion	+	78	625
C08389	Abortion	+	1250	7500
G03652	Meningitis	+	1875	1875
LM4	Milk	+	1875	6250
LM6	Milk	+	1875	2500
LM7	Milk	+	1875	5000

[a] All strains showed equivalent levels of weak haemolysis on sheep blood agar.
[b] MIC = minimum inhibitory concentration.
[c] MBC = minimum bactericidal concentration.

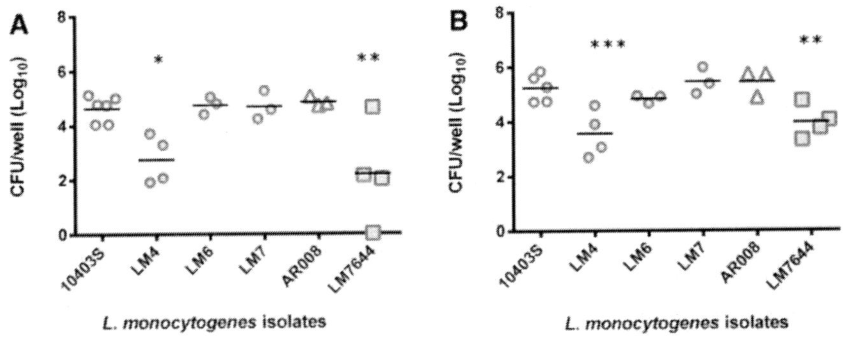

Figure 2. Infection of Caco2 cells. Caco2 cells were infected with an MOI = 25 (A) or 200 (B) for 2 h. For each strain a minimum of 3 independent experiments were performed. 10403S $n = 6$, LM4 $n = 4$, LM6 $n = 3$, LM7 $n = 3$, AR008 $n = 3$, LM7644 $n = 4$. All data points and mean are shown. *$p < 0.05$; **$p < 0.01$; ***$p < 0.001$ compared to 10403S (One way ANOVA & Tukey's multiple comparison test).

3.5. Sensitivity to lysozyme

Since lysozyme is one of the main antibacterial defence mechanisms in the eye, we performed MIC/MBC assays to determine the lysozyme sensitivity of our isolates. The healthy eye isolate AR008 and the abortion strain LM7644, which were both less able to infect the conjunctiva explants, showed the lowest resistance towards lysozyme treatment (MIC = 78 μg ml^{-1}, Table 2). While the MBC value for LM7644 was also low (625 μg ml^{-1}), that for AR008 was in the same range as the lysozyme-resistant strains (Table 2).

A survival assay was also performed using the highest concentration of lysozyme reported in the literature in bovine eyes (580 μg ml^{-1}) (Gionfriddo et al., 2000). This experiment confirmed that both AR008 and LM7644 were the most sensitive to lysozyme challenge and that they were significantly more sensitive than all the bovine conjunctivitis isolates (Fig. 3). Two other strains, the abortion isolate C08389 and the meningitis isolate G03652, also showed lower levels of lysozyme resistance than the majority of the isolates tested (Fig. 3).

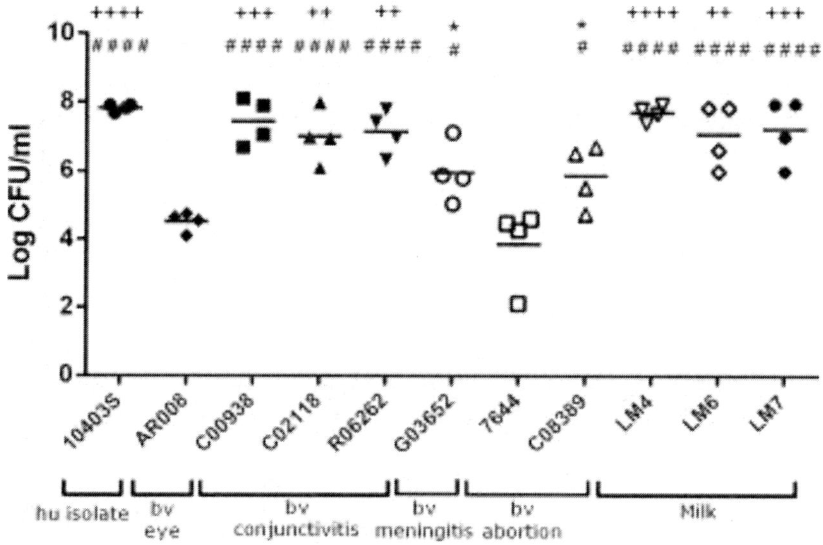

Figure 3. Lysozyme sensitivity of *L. monocytogenes* isolates. Bacteria were incubated with 580 µg ml^{-1}lysozyme for 24 h at 37 °C ($n = 4$). All data points and mean are shown. *$p < 0.05$; ****$p < 0.0001$ compared to LM10403S; $^{++}p < 0.01$; $^{+++}p < 0.001$, $^{++++}p < 0.0001$compared to AR008; $^{\#}p < 0.05$;$^{\#\#\#\#}p < 0.0001$ compared to LM7644 (One way ANOVA and Tukey's multiple comparison test).

4. DISCUSSION

From a total of 46 eye swabs, either from cattle with 'weepy' eyes or clinically healthy eyes post-slaughter, *L. monocytogenes* was only recovered from two samples, neither of which had any sign of clinical infection. Given the limit of detection of the swabbing technique was determine in the laboratory to be 10^3 cells per eye, this suggests that *L. monocytogenes* can be found at reasonably high levels in cattle eyes without causing infection.

The two strains that were least able to infect the conjunctiva were also the stains that were most sensitive to lysozyme treatment, indicating that perhaps the ability of *L. monocytogenes* isolates to infect bovine conjunctiva is associated with lysozyme resistance. While reports of lysozyme concentrations found within the bovine eye are variable (0–580 µg ml^{-1}) (Gionfriddo et al., 2000 and Prieur, 1986), our results using the upper range of these values suggest that there is a relationship between the ability to infect explanted conjunctiva tissues and natural levels of resistance to lysozyme.

In *Listeria*, modification of the peptidoglycan (PG) by PdgA and OatA has been shown to make the cells resistant to lysozyme by modifying the sugar backbone so that it is a poor substrate for the enzyme (Bera et al., 2005 and Vollmer and Tomasz, 2000). These modifications are known to be important for survival inside the mammalian host, and*pgdA* and *oatA* mutants are also both impaired in their ability to survive in macrophages (Boneca et al., 2007 and Rae et al., 2011). Recently it has been suggested that natural levels of lysozyme resistance of *L. monocytogenes* are due to the up-regulation of PG modifying-enzymes rather than acquisition of novel PG modifications, through the actions of the orphan response regulator DegU and an abundant non-coding RNA encoded by*rli31* (Burke et al., 2014). Mutations in both of these regulatory elements resulted in lysozyme sensitivity and reduced virulence in mice. In contrast the invasion of Caco2 cells, a non-phagocytic epithelial cell line, was not impaired in *L. monocytogenes* lacking PgdA, suggesting that this peptidoglycan modification which confers lysozyme resistance is not essential to be able to enter and survive within non-phagocytic cells (Boneca et al., 2007). This is consistent with our observation that the lysozyme-sensitive strain AR008 showed no impairment in its ability to infect Caco2 cells. However, the bovine abortion strain LM7644 was less able to infect Caco2 cells, suggesting a different basis for the lysozyme sensitivity of this strain.

Here we present the first bovine conjunctiva explant model for infection studies and demonstrate that clinical *L. monocytogenes* isolates from cases of bovine keratoconjunctivitis are able to infect these tissues recovered from animals slaughtered for commercial use. We have used this model to show that lysozyme resistance may be a key property to establish eye infections, and that cattle eyes may be colonised, but not infected by, strains of *Listeria* with naturally lower levels of lysozyme resistance. This model will allow us to study the host pathogen interactions that occur during the establishment of eye infections without the requirement for the use of live animals which is in keeping with the drive towards reduced animal usage in studies of infection. In addition, this model could be used to study other infectious bacteria which commonly cause conjunctivitis in cattle, such as *Moraxella* spp. (Henson and Grumbles, 1960).

CONFLICTS OF INTEREST

Authors declare that they have no conflicts of interest.

ACKNOWLEDGEMENTS

This work was funded by the University of Nottingham; and JW and AF were awarded BBSRC Research Experience Placements.

REFERENCES

1. Aubry, C., Goulard, C., Nahori, M.A., Cayet, N., Decalf, J., Sachse, M., Boneca, I.G., Cossart, P., Dussurget, O., 2011. OatA, a peptidoglycan O-acetyltransferase involved in Listeria monocytogenes immune escape, is critical for virulence. J. Infect. Dis. 204, 731–740.

2. Aubry, C., Corr, S.C., Wienerroither, S., Goulard, C., Jones, R., Jamieson, A.M., Decker, T., O'Neill, L.A., Dussurget, O., Cossart, P., 2012. Both TLR2 and TRIF contribute to interferon-beta production during Listeria infection. PLoS ONE 7, e33299.

3. Bera, A., Herbert, S., Jakob, A., Vollmer, W., Gotz, F., 2005. Why are pathogenic staphylococci so lysozyme resistant? The peptidoglycan O-acetyltransferase OatA is the major determinant for lysozyme resistance of Staphylococcus aureus. Mol. Microbiol. 55, 778–787.

4. Bishop, D.K., Hinrichs, D.J., 1987. Adoptive transfer of immunity to Listeria monocytogenes. The influence of in vitro stimulation on lymphocyte subset requirements. J. Immunol. 139, 2005–2009.

5. Boneca, I.G., Dussurget, O., Cabanes, D., Nahori, M.A., Sousa, S., Lecuit, M., Psylinakis, E., Bouriotis, V., Hugot, J.P., Giovannini,M., Coyle, A., Bertin, J., Namane, A., Rousselle, J.C., Cayet, N., Prevost, M.C., Balloy, V., Chignard, M., Philpott, D.J., Cossart, P., Girardin, S.E., 2007. A critical role for peptidoglycan N-deacetylation in

Listeria evasion from the host innate immune system. Proc. Natl. Acad. Sci. U. S. A. 104, 997–1002.

6. Bundrant, B.N., Hutchins, T., den Bakker, H.C., Fortes, E., Wiedmann, M., 2011. Listeriosis outbreak in dairy cattle caused by an unusual Listeria monocytogenes serotype 4b strain. J. Vet. Diagn. Investig. 23, 155–158.

7. Burke, T.P., Loukitcheva, A., Zemansky, J., Wheeler, R., Boneca, I.G., Portnoy, D.A., 2014. Listeria monocytogenes is resistant to lysozyme through the regulation, not the acquisition, of cell wall-modifying enzymes. J. Bacteriol. 196, 3756–3767.

8. Cossart, P., Toledo-Arana, A., 2008. Listeria monocytogenes, a unique model in infection biology: an overview. Microb. Infect. 10, 1041–1050.

9. Doughty, M.J., Hagan, S., 2013. On the staining of human tissue-cultured (Chang) conjunctival cells with rose bengal and lissamine green. Cont. Lens Anterior Eye 36, 32–40.

10. Doumith, M., Buchrieser, C., Glaser, P., Jacquet, C., Martin, P., 2004. Differentiation of the major Listeria monocytogenes serovars by multiplex PCR. J. Clin. Microbiol. 42, 3819–3822.

11. Erdogan, H.M., 2010. Listerial keratoconjunctivitis and uveitis (silage eye). Vet. Clin. North Am. Food Anim. Pract. 26, 505–510.

12. Erdogan, H.M., Cetinkaya, B., Green, L.E., Green, L.E., Cripps, P.J., Morgan, K.L., 2001. Prevalence, incidence, signs and treatment of clinical listeriosis in dairy cattle in England. Vet. Rec. 149, 289–293.

13. Gaillard, J.L., Berche, P., Mounier, J., Richard, S., Sansonetti, P., 1987. In vitro model of penetration and intracellular growth of Listeria monocytogenes in the human enterocyte-like cell line Caco-2. Infect. Immun. 55, 2822–2829.

14. Gionfriddo, J.R., Davidson, H., Asem, E.K., Krohne, S.G., 2000. Detection of lysozyme in llama, sheep, and cattle tears. Am. J. Vet. Res. 61, 1294–1297.

15. Henson, J.B., Grumbles, L.C., 1960. Infectious bovine keratoconjunctivitis. I. Etiology. Am. J. Vet. Res. 21, 761–766.

16. Kuhbacher, A., Cossart, P., Pizarro-Cerda, J., 2014. Internalization assays for Listeria monocytogenes. Methods Mol. Biol. 1157, 167–178.

17. Lawrence, L.H., Hill, H.A., Linton, M., 1995. DNA sequence divergence within the iap gene of L. monocytogenes as a means of characterising strains.In: XII International Symposium on Problems of Listeriosis, Perth, Australia.

18. McLauchlin, J.R., Rees, C.E.D., Dodd, C.E.R., 2014. Listeria monocytogenes and the Genus isteria. In: Rosenberg, E., DeLong, E.F., Lory, S., Stackebrandt, E., Thompson, F. (Eds.), The Prokaryotes: Firmicutes and Tenericutes. Springer, pp. 241–259.

19. Portnoy, D.A., Chakraborty, T., Goebel, W., Cossart, P., 1992. Molecular determinants of Listeria monocytogenes pathogenesis. Infect. Immun. 60, 1263–1267.

20. Prieur, D.J., 1986. Tissue specific deficiency of lysozyme in ruminants. Comp. Biochem. Physiol. B: Comp. Biochem. 85, 349–353.

21. Rae, C.S., Geissler, A., Adamson, P.C., Portnoy, D.A., 2011. Mutations of the Listeria monocytogenes peptidoglycan N-deacetylase and O-acetylase result in enhanced lysozyme sensitivity, bacteriolysis, and hyperinduction of innate immune pathways. Infect. Immun. 79, 3596–3606.

22. Revathi, R., Bhat, P., Saraf, A., 2011. Bacterial conjunctivitis and keratitis. In: Vanathi, Z.C.A.M. (Ed.), Postgraduate Ophthalmology. Jaypee, pp. 495–506.

23. Sherman, M.D., Cevallos, V., Gabriel, R., Dawson, C., Stephens, R.S., 1994. Measurement of human tear lysozyme using a novel synthetic substrate. Adv. Exp. Med. Biol. 350, 363–366.

24. Vollmer, W., Tomasz, A., 2000. The pgdA gene encodes for a peptidoglycan N-acetylglucosamine deacetylase in Streptococcus pneumoniae. J. Biol. Chem. 275, 20496–20501.

25. Ward, T.J., Gorski, L., Borucki, M.K., Mandrell, R.E., Hutchins, J., Pupedis, K., 2004. Intraspecific phylogeny and lineage group identification based on the prfA virulence gene cluster of Listeria monocytogenes. J. Bacteriol. 186, 4994–5002.

26. Wiedmann, M., Bruce, J.L., Knorr, R., Bodis, M., Cole, E.M., McDowell, C.I., McDonough, P.L., Batt, C.A., 1996. Ribotype diversity of Listeria monocytogenes strains associated with outbreaks of listeriosis in ruminants. J. Clin. Microbiol. 34, 1086–1090

CHAPTER 11

Sample Size Considerations for One-to-One Animal Transmission Studies of the Influenza A Viruses

Hiroshi Nishiura[1,2]*, Hui-Ling Yen[3] , Benjamin J. Cowling[1]

1 School of Public Health, The University of Hong Kong,
Hong Kong SAR, China

2 PRESTO, Japan Science and Technology Agency, Saitama, Japan

3 Centre of Influenza Research, School of Public Health, Li Ka Shing Faculty of Medicine, The University of Hong Kong,
Hong Kong SAR, China

ABSTRACT

Background

Animal transmission studies can provide important insights into host, viral and environmental factors affecting transmission of viruses including influenza A. The basic unit of analysis in typical animal transmission experiments is the presence or absence of transmission from an infectious animal to a susceptible animal. In studies comparing two groups (e.g. two host genetic variants, two virus strains, or two arrangements of animal cages), differences between groups are evaluated by comparing the proportion of pairs with successful transmission in each group. The present study aimed to discuss the significance and power to estimate transmissibility and identify differences in the transmissibility based on one-to-one trials. The analyses are illustrated on transmission studies of influenza A viruses in the ferret model.

Methodology/Principal Findings

Employing the stochastic general epidemic model, the basic reproduction number, R_0, is derived from the final state of an epidemic and is related to the probability of successful transmission during each one-to-one trial. In studies to estimate transmissibility, we show that 3 pairs of infectious/susceptible animals cannot demonstrate a significantly higher transmissibility than $R_0=1$, even if infection occurs in all three pairs. In comparisons between two groups, at least 4 pairs of infectious/ susceptible animals are required in each group to ensure high power to identify significant differences in transmissibility between the groups.

Conclusions

These results inform the appropriate sample sizes for animal transmission experiments, while relating the observed proportion of infected pairs to R_0, an interpretable epidemiological measure of transmissibility. In addition to the hypothesis testing results, the wide confidence intervals of R_0 with small sample sizes also imply that the objective demonstration of difference or similarity should rest on firmly calculated sample size.

1. INTRODUCTION

The transmission potential of a respiratory virus is commonly measured by the basic reproduction number, R_0, i.e. the average number of secondary cases produced by a typical primary case throughout the entire course of host infection, which has been regarded as one of the most important quantities in infectious disease epidemiology [1]. The value of R_0not only informs how transmissible an infected individual is, but also gives three epidemiological insights into the transmission dynamics, i.e., (i) the risk of observing an epidemic given a certain number of infected individuals, (ii) the risk of infection in an individual throughout the course of an epidemic (given the epidemic), and (iii) the minimum control effort that is required to prevent or curb an epidemic based on a threshold theorem[1]. Among various methods for estimating R_0, animal transmission experiments have been used as a useful tool for measuring the transmissibility in controlled conditions [2], although it should be noted that R_0 is not only the property of a pathogen but also reflects the specific entire host-pathogen-environmental system, and thus, R_0 in the experimental setting is not directly applicable to other (e.g. natural) settings. However, through transmission experiments, one can identify the importance of various factors on transmissibility, including host (e.g. genetic variants or immune status), virus (e.g. different strains), and environmental (e.g. ambient temperature/humidity) factors and their interactions, thereby providing important insights into mechanisms of infection and transmission.

Transmission experiments of influenza A viruses have helped to determine the molecular mechanisms of adaptation in human host due to the multi-host nature of the virus [3]–[5]. It is possible to study transmission in humans in controlled experimental settings [6] and natural settings [7], but such studies are resource intensive. Moreover, prior influenza exposure history in humans varies between individuals and is difficult to control. Although animal studies cannot replace human studies, they can provide complementary information on factors affecting transmission. The ferret model tends to be preferred over the mouse and guinea pig models for transmission experiments, because ferrets display clinical signs and symptoms which include fever, nasal discharge, and sneeze reflex [8]–[10]. In some studies, the potential of respiratory droplet transmission (inclusive of both droplet and aerosol transmissions) has been examined by placing a susceptible ferret in a cage next to another cage with a ferret inoculated with an influenza virus and allowing an exchange of air between the cages. Typically, such one-to-one transmission has been examined for two to four pairs for each virus, and the proportion of pairs with successful transmission has been compared between two (or more) different influenza viruses.

Despite a number of published ferret transmission experiment studies of influenza viruses, there has been no explicit estimation of R_0 from experiments that typically involve very small sample sizes, and moreover, the sample size rationale of such an experiment has not been extensively discussed. Although a common approach is now to use three pairs of ferrets in each virus group, and to compare the proportions of pairs with successful transmission between each group, the appropriate sample size and proper analysis of results have not been investigated. For instance, suppose that researchers did not observe any transmission for one virus (i.e. $k/n=0/3$ where n and k represent the numbers of pairs and infected pairs, respectively), while all three pairs resulted in transmission for the other virus (i.e. $k/n=3/3$). Given this result, we would like to know (i) if the difference in the transmissibility between the two groups is statistically significant and (ii) the degree of difference in transmissibility given the small sample size. The present study aimed to discuss the significance and power of one-to-one animal transmission experiments with particular reference to ferret transmission studies of influenza A viruses as a case study.

2. MATERIALS AND METHODS

2.1 One-to-one transmission experiment data

We start by presenting summary results of the published transmission experiment studies of influenza A viruses using the ferret model. Table 1 summarizes a total of 12 transmission experiment studies that were conducted under a one-to-one transmission experiment design [3], [5], [8], [11]–[19]. In the present study, we restrict our interest to the ferret model, especially its use in examining respiratory transmission, for consistency and clarity both in theory and biology. Among the total of 12 studies, nine investigated H1N1 viruses including two on 1918–19 pandemic viruses and seven on 2009 pandemic viruses. Three studies investigated the transmissibility of H5N1 viruses, two on H3N2 viruses and one on H2N2 viruses. In principle, those studies share the experimental design (i.e. inoculation of one ferret in a cage and exposure of the other ferret in an adjacent cage), but the details have been variable. The air flow (e.g. direction and air exchange rate) has not been strictly regulated by common rules, and viral dose for inoculation have not been identical among these studies, and thus, various differences in experimental designs prohibit pooling of data from differently designed studies.

Table 1. Study characteristics of the published one-to-one animal transmission experiment of influenza A viruses.

	Year of publication	Subtype	Infected animals in control group (k/n)†	Infected animals in comparison group (k/n)†	Judgment‡	Study objective(s)
Maines et al. [11]	2006	H3N2 & H5N1	3/3	0/3	Different	To identify the importance of virus internal protein genes in regulating transmissibility
Tumpey et al. [12]	2007	H1N1	3/3	0/3	Different	To show that only a modest change in the 1918 influenza hemagglutinin receptor binding site alters the transmissibility
van Hoeven et al. [8]	2009	H1N1	3/3	0/3	Different	To identify genetic determinants that govern airborne transmission among 1918-avian H1N1 ifnluenza reassortant viruses
Munster et al. [13]	2009	H1N1	4/4	4/4	Similar	To study aerosol transmission of the pandemic 2009 A(H1N1) influenza virus as compared with a seasonal 2007 A(H1N1) virus
Maines et al. [14]	2009	H1N1	3/3	2/3	Different	To compare the transmissibility of H1N1-2009 against a seasonal H1N1 virus through respiratory droplets
Pappas et al. [15]	2010	H2N2	3/3	0/3	Different	To examine the transmissibility of human H2N2 viruses isolated during the 1957/58 pandemic
Kiso et al. [16]	2010	H1N1	3/3	3/3	Similar	To compare the transmissibility between oseltamivir-resistant and sensitive H1N1-2009.

Van Doremalen et al. [17]	2011	H1N1	3/3	3/3	Similar	To investigate the effect of residue 227 in hemagglutinin on cell tropism and transmission of pH1N1 2009.
Koster et al. [18]	2012	H1N1	4/4	4/4	Similar	To develop a method to detect exhaled viral aerosol transmission between unanesthetized infected and susceptible ferrets.
Pearce et al. [19]	2012	H3N2	3/3	3/3	Similar	To analyze the transmissibility of four A(H3N2)v influenza viruses isolated from humans in 2009, 2010 and 2011.
Herfst et al. [5]	2012	H5N1 & H1N1	0/2	3/4	Different	To address the concern that the virus could acquire the ability of airborne transmission under natural conditions
Imai et al. [3]	2012	H5N1 & H1N1	0/3	4/6	Different	To assess the molecular changes in HA that would allow a virus possessing subtype H5 HA to be transmissible

[*] *n* represents the sample size, i.e. the number of pairs of a single infected animal and a susceptible animal. *k* represents the number of pairs with successful transmission. The numbers among control group are the highest reported frequency of infected animals in control group, while the numbers among comparison group show the lowest reported frequency of infection. [‡]Judgment corresponds to the interpretation of difference in transmissibility.

With regard to the sample size (i.e. the number of pairs for each virus), all studies used small numbers of ferrets, with most choosing 3 pairs of infectious/susceptible ferrets in each group. There were two studies that used 4 pairs. Two studies used different pair numbers between two groups, i.e., one study used 2 pairs for the control and 4 pairs for the comparison group, and the other study used 3 pairs for the control and 6 pairs for the comparison group. The results (i.e. the number of pairs with successful transmission) shown in Table 1 represent the highest and lowest reported numbers among the all combinations of two viruses, and the judgment of difference (or similarity) was drawn in the original publications based on the corresponding results. The primary objective of the original studies was either to identify molecular mechanisms (e.g. specific viral gene, amino acid or protein) governing the transmissibility (n=5) or to quantify or compare the capacity of aerosol transmissibility (n=7).

2.2 Stochastic general epidemic model

To allow comparison of the transmissibility, here we express the result from one-to-one transmission experiment (i.e. the proportion of pairs with transmission) as a function of R_0. First of all, we adopt an assumption that each pair is independent of other pairs, including no air-exchange between pairs. Let $p_{s,i}(t)$ be the conditional probability of observing $s$susceptible and i infected ferrets at time t given the initial condition of susceptible and infected ferrets (s_0, i_0) at time 0, i.e.

$$p_{s,i}(t) = \Pr\{(S(t),I(t))=(s,i)|(S(0),I(0))=(s_0,i_0)\}, \tag{1}$$

then the so-called "stochastic general epidemic" model is described by the following differential-difference equation:

$$\frac{d}{dt}p_{s,i}(t)=(i+1)p_{s,i+1}(t)+\frac{R_0}{N-1}(s+1)(i-1)p_{s+1,i-1}(t) - \frac{R_0}{N-1}sip_{s,i}(t)-ip_{s,i}(t), \tag{2}$$

where N is the total population size (N=2 in the case of one-to-one experiment) and t represents the multiple of the mean infectious period (i.e. the time unit is normalized by the mean infectious period). Here it should be noted that the mass action part has been scaled by (N−1), and not by N, because of small population size that requires us to precisely consider the impact of N on the incidence term. That is, in the case of small N, the transient risk of infection should be proportional to I(t)/(N−1) which can be exemplarily understood for N=2 (i.e. if we use I(t)/N for calculating the incidence, it would indicate erroneously that the half of infectious individuals I(t)/2 would contribute to the transmission). Since the initial condition gives $p_{s0,i0}(0)=1$, the probability of successful transmission by infinite time, q, is computed by $p_{0,0}(\infty)$ [20]–[22] and the solution is

$$q=\frac{R_0}{R_0+1}. \tag{3}$$

Note that the analytical solution of q for small N is $q=R_0/(R_0+N-1)$ which is different from what has been previously discussed [22]. Since the one-to-one transmission experiment handles the binary outcome (i.e. success or failure of transmission), the probability of transmission is computed by employing a binomial distribution. That is, for n independent pairs of one-to-one transmission experiments, the probability of observing k pairs with successful transmission is

$$\Pr(K=k;n,R_0)=\binom{n}{k}\left(\frac{R_0}{R_0+1}\right)^k\left(\frac{1}{R_0+1}\right)^{n-k}. \tag{4}$$

The maximum likelihood estimator of R_0 based on the observed average frequency of successful transmission, k/n, is given by equating

$$\hat{R}_0=\frac{k}{n-k}. \tag{5}$$

The 100(1-α)% confidence interval of R_0 is calculated from the solution of $R_{0,CI}=x/(n-x)$ in which x satisfies

$$\sum_{i=x}^{n}\binom{n}{i}\left(\frac{\hat{R}_0}{\hat{R}_0+1}\right)^i\left(\frac{1}{\hat{R}_0+1}\right)^{n-i}=\frac{\alpha}{2}, \tag{6}$$

for the upper bound and

$$\sum_{i=0}^{x}\binom{n}{i}\left(\frac{\hat{R}_0}{\hat{R}_0+1}\right)^i\left(\frac{1}{\hat{R}_0+1}\right)^{n-i}=\frac{\alpha}{2}, \tag{7}$$

for the lower bound, except that the lower bound is 0 when x=0 and the upper bound is infinity when x=n. In these exceptional circumstances, the upper bound for x=0 is calculated as $R_{0,upper}=((\alpha/2)^{(1/n)}-1)/(\alpha/2)^{(1/n)}$ and the lower bound for x=n is calculated as $R_{0,lower}=(\alpha/2)^{(1/n)}/(1-(\alpha/2)^{(1/n)})$ as can be derived from the binomial distribution [23]. In addition to the final size discussed above, statistical consideration of transient state has been given elsewhere [24].

2.3 Hypothesis testing

We subsequently consider the difference in the transmissibility in published experimental studies in two different ways, because the null hypothesis has not necessarily been mentioned in the original articles in Table 1. Let $R_{0,ref}$ be a specified reference value of the basic reproduction number. The first possible way to compare the transmissibility is to regard the result from each virus as one-sample comparison, which may be the case when$R_{0,ref}$ of control virus can be assumed known (e.g. pre-determined) from published studies and so on. In this scenario, we compare R_0 against $R_{0,ref}$, i.e.

$$H_0 : R_0 \le R_{0,ref} \text{ vs. } H_a : R_0 > R_{0,ref}, \quad \textbf{(8)}$$

which may sometimes be intended to support the notion that some key molecular structure helped to acquire substantial transmissibility for a specific virus (e.g. by setting $R_{0,ref}=1$ or$R_{0,ref}=0$). It should be noted that R_0 depends on experimental design (air change rate per hour, air flow direction, etc) and is not comparable between differently designed experiments. Using the relationship (5), the issue of comparing transmissibility is replaced by one-sample comparison of a binomial proportion. The p-value for testing (8) given that $k$or more pairs resulted in infection is computed by

$$\Pr(X \ge k; n, R_{0,ref}) = \sum_{i=k}^{n} \binom{n}{i} \left(\frac{R_{0,ref}}{R_{0,ref}+1}\right)^{i} \left(\frac{1}{R_{0,ref}+1}\right)^{n-i}, \quad \textbf{(9)}$$

which rejects H_0 if less than or equal to $\alpha=0.05$. Demonstrating $R_0>R_{0,ref}=0$ would help researchers to demonstrate the capacity of a ferret infected with a virus to “cause” (any number of) secondary transmission. Demonstrating $R_0>R_{0,ref}=1$ is particularly useful in practice, because satisfying this condition indicates that, at least under experimental conditions, a single primary case can on average generate one or more secondary cases through the respiratory route (and thus, is regarded as possessing a substantial transmission potential to cause an epidemic only through that particular mode of transmission). Of course, rejecting H_0: $R_0 \geq 1$ can be achieved similarly by calculating $\Pr(X<k;\ n, R_{0,ref}=1) \leq 0.05$. The power for (8) is given by

$$\sum_{k=0}^{n} \binom{n}{k} \left(\frac{R_0}{R_0+1}\right)^{k} \left(\frac{1}{R_0+1}\right)^{n-k} I(\Pr(X \geq k; n, R_{0,ref}) \leq \alpha), \quad (10)$$

where $I(.)$ is the indicator function.

The second way to test the transmissibility is to consider the two virus groups within the same study in Table 1 as the comparison of two binomial proportions, q_1 and q_2 (or equivalently the comparison of two basic reproduction numbers estimated for respective viruses, $R_{0,1}$ and $R_{0,2}$) under the hypotheses

$$H_0 : q_1 = q_2 \text{ vs. } H_a : q_1 \neq q_2, \quad (11)$$

(or H_0: $R_{0,1}=R_{0,2}$) and the implementation is exactly the same as two-tailed exact test for two samples that has been already discussed elsewhere [25]. It should be noted that differing number of pairs between two virus groups can be easily addressed by varying sample sizes in the exact test. In both one-sample and two-sample cases, the sample size estimation would have to be made directly from the binomial distribution (e.g. from (10) with a desired power). However, as an alternative, the power calculation could rest on a modified Wald test statistic, i.e., the well-known score confidence interval proposed by Agresti and Coull [26],[27], [28].

For numerical illustrations, we consider the p-value and power for all possible patterns of final size for the number of pairs, $n=3$, 4 and 5 for both one-sample and two-sample comparisons. These numbers of pairs are specifically considered, because 3 pairs have been conventionally adopted, and we anticipate that 6 or more pairs may not be logistically very feasible for testing many types of influenza virus at present. A one-sample comparison is made by a one-tailed Fisher's exact test, while a two-sample comparison rests on a two-tailed test. While restricting our consideration to $n=3$, 4 and 5, we also examine the p-value for one-sample test with varying reference values of the basic reproduction number and the number of pairs (from 1 to 10), especially in the case we have $k=n$ (i.e. all pairs resulted in infection) or $k=n-1$ which are frequently the case in published experimental studies.

3. RESULTS

3.1 One-sample comparison

Table 2 shows the p-value and power for one-sample comparisons given that the number of pairs was 3, 4 or 5. Even when all pairs result in infection during 3 or 4 pair study, the experiment cannot indicate that the R_0 is significantly greater than 1. Only when we have a result of 5/5, the difference can be stated as significantly greater than R_0 of 1. Moreover, in the cases all pairs result in infection, one can quantitatively examine only the lower bound of R_0, and the expected value and the upper bound of R_0 are calculated as infinite. Figure 1shows the p-value with different numbers of pairs in the case that all pairs are infected (i.e. $k=n$) or all pairs minus 1 resulted in infection (i.e. $k=n-1$). When the reference value of R_0 is as large as 2, three pairs are not large enough to demonstrate $R_0>2$ at a significant level $\alpha=0.05$, and one may need at least 8 pairs and all the eight pairs need to be infected. At a stricter threshold (e.g. $\alpha=0.01$), seven or more pairs would be required to reject $R_{0,\text{null}}=1$. In the case of $k=n-1$, even ten pairs would not be enough to demonstrate that $R_0>2$ at $\alpha=0.05$ given $k=n-1$.

The p-values are shown to indicate the significance level at which the estimated basic reproduction number is significantly greater than the null value ($R_{0,\text{null}}$ are set to be 0.2, 1.0 and 2.0 with the null hypothesis, H_0: $R_0 \leq R_{0,\text{null}}$) given that all pairs resulted in infection (black lines; $n=k$ where n and k are the numbers of pairs and infected pairs, respectively) or all pairs minus 1 resulted in infection (grey lines; $n=k+1$). The hypothesis testing is based on one-sample comparison using the one-tailed Fisher's exact test. The horizontal grey bold line represents the significance level at 0.05.

Also, when one pair escapes infection (i.e. $k=n-1$), Figure 1 and Table 2 consistently suggest that a five-pair or smaller study cannot help judge if R_0 is significantly greater than 1. The results 2/3, 3/4 or 4/5 does not indicate significant difference from $R_0=1$. All other combinations in Table 2 cannot determine if R_0 is significantly greater than 1, and more importantly, either lower or upper 95% confidence interval of estimated R_0 for these combinations always takes an extreme value (i.e. either lower bound being 0 or upper bound being infinite).

Provided that a transmission study intends to demonstrate $R_0>0$, the presence of at least one successful transmission (i.e. any k/n except for $k=0$) can yield a significant result with $p<0.01$. However, it should be remembered that power is not substantial for $k/n=1/4$ and $1/5$ (Table 2). When one intends to demonstrate that the transmission potential is less than 1, the examined total sample sizes are not enough to argue significant differences (Table 2). That is, given the number of pairs is 3, 4 or 5, it is more feasible to show that $R_0>1$ than demonstrating $R_0<1$.

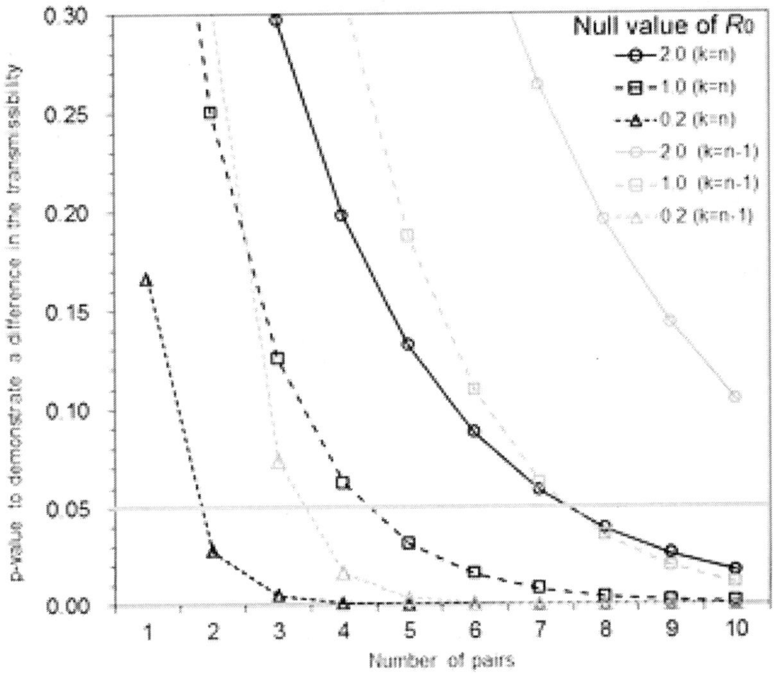

Figure 1. The p-value to demonstrate the significant difference in the transmissibility based on fully or nearly fully successful one-to-one transmission experiment.

3.2 Two-sample comparison

Table 3 summarizes the p-value and power for two-sample comparisons given that the number of pairs was 3, 4 or 5. Given three pairs for each sample, it is impossible to demonstrate any significant difference between two sample groups. In the case of four pairs, only a combination of 0/4 and 4/4 can indicate a significant difference. Given five pairs, three combinations (i.e. 0/5 vs 5/5, 1/5 vs 5/5 or 0/5 vs 4/5) could suggest significant difference in the transmission potential.

Table 2. One-tailed test results of the basic reproduction number based on one-to-one transmission experiment.

Number of pairs	Final size	R_0[†]	(95% CI[‡])	p-value to demonstrate $R_0>1$	power to demonstrate $R_0>1$	power to demonstrate $R_0>0$	p-value to demonstrate $R_0<1$
3	0	0.0	(0, 4.8)	1.00	0.00	1.00	1.00
	1	0.5	(0, ∞)	0.88	0.00	1.00	0.74
	2	2.0	(0, ∞)	0.50	0.00	1.00	0.70
	3	∞	(0.8, ∞)	0.13	0.00	1.00	1.00
4	0	0.0	(0, 3.0)	1.00	0.00	0.00	1.00
	1	0.3	(0, 6.0)	0.94	0.00	0.68	0.74
	2	1.0	(0, ∞)	0.69	0.00	0.94	0.69
	3	3.0	(0.7, ∞)	0.31	0.00	0.99	0.68
	4	∞	(1.3, ∞)	0.06	0.00	1.00	1.00
5	0	0.0	(0, 2.2)	1.00	0.00	0.00	1.00
	1	0.3	(0, 3.0)	0.97	0.00	0.67	0.74
	2	0.7	(0, 8.0)	0.81	0.00	0.92	0.68
	3	1.5	(0.5, ∞)	0.50	0.08	0.99	0.66
	4	4.0	(1.3, ∞)	0.19	0.33	1.00	0.67
	5	∞	(1.8, ∞)	0.03[¶]	1.00	1.00	1.00

[†]The basic reproduction number, estimated from the one-to-one transmission experiment. [‡]CI, confidence intervals. [¶]R_0 is significantly greater than 1 by one-sample Fisher's exact test.

Table 3. Two-tailed comparison of the basic reproduction numbers based on one-to-one transmission experiment (H_0: $R_0 = R_{0,null}$[†]).

Number of pairs	Final size of comparison group	0		1		2		3		4		5	
		p-value	power	p-value	power	p-value	power	p-value	power	p-value	power	p-value	power
3	0	1.00	0.00	1.00	0.04	0.40	0.30	0.10	1.00				
	1	1.00	0.04	1.00	0.00	1.00	0.09	0.40	0.30				
	2	0.40	0.30	1.00	0.09	1.00	0.00	1.00	0.04				
	3	0.10	1.00	0.40	0.30	1.00	0.04	1.00	0.00				
4	0	1.00	0.00	1.00	0.05	0.43	0.31	0.14	0.74	0.03[¶]	1.00		
	1	1.00	0.05	1.00	0.04	1.00	0.13	0.49	0.37	0.14	0.74		
	2	0.43	0.31	1.00	0.13	1.00	0.07	1.00	0.13	0.43	0.31		
	3	0.14	0.74	0.49	0.37	1.00	0.13	1.00	0.04	1.00	0.05		
	4	0.03[¶]	1.00	0.14	0.74	0.43	0.31	1.00	0.05	1.00	0.00		
5	0	1.00	0.00	1.00	0.01	0.44	0.09	0.17	0.34	0.05[¶]	0.74	0.01[¶]	1.00
	1	1.00	0.01	1.00	0.01	1.00	0.03	0.52	0.14	0.29	0.38	0.05[¶]	0.74
	2	0.44	0.09	1.00	0.03	1.00	0.02	1.00	0.05	0.52	0.14	0.17	0.34
	3	0.17	0.34	0.52	0.14	1.00	0.05	1.00	0.02	1.00	0.03	0.44	0.09
	4	0.05[¶]	0.74	0.21	0.38	0.52	0.14	1.00	0.03	1.00	0.01	1.00	0.01
	5	0.01[¶]	1.00	0.05[¶]	0.74	0.17	0.34	0.44	0.09	1.00	0.01	1.00	0.00

[†] $R_{0,null}$, the basic reproduction number to be used in null hypothesis. [¶] Estimated transmissibility from one-to-one transmission experiment is significantly different from $R_{0,null}$ by two-sample Fisher's exact test.

4. DISCUSSION

The present study discussed the sample size considerations for one-to-one experimental studies of the transmission of influenza A viruses. Employing the stochastic general epidemic model, R_0 was derived from the final state of an epidemic [29], [30], [31], and its relevance to the probability of successful transmission during the one-to-one trial was explained. Three findings are particularly notable. First, k/n=3/3 and 4/4 are not indicative of significant excess of R_0 from 1 in one-sample comparison. At least, five pairs would be required to demonstrate significant difference in the one-sample comparison. Second, n=3 is not enough to show any significant difference in two-sample comparisons. Third, $k=n$ can yield the significant difference when n=5 or greater in one-sample comparison, but one has to remember that the expected value and the upper confidence interval of R_0 would be calculated as infinite for small n. That is, while the experiment may be able to show significant difference from reference value, $k=n$ can inform only the lower bound of R_0. With the very limited sample sizes such as n=3, 4 or 5, it is always the case that either lower or upper 95% confidence interval takes an extreme value (i.e. lower =0 or upper =∞). Keeping these points in mind, one can plan the one-to-one transmission study with reference to our computed results in Tables 2 and 3, while relating the observed proportion of infected pairs to R_0, an interpretable epidemiological measure of transmissibility.

The most important caveat in the present study in relation to the common practice is that n=3 is not enough to show a significant difference as well as $R_0>1$ for one-sample comparison, while it can demonstrate $R_0>0$. Moreover, comparing a group with n=3 against a reference group with the same sample size does not allow researchers to demonstrate any significant difference in the transmissibility between two sample groups. If two samples have to be compared, n=4 would be regarded as minimum, and moreover, $k=n$ for n=4 and $k=n$ or $k=n-1$ for n=5, respectively, would have to be required along with the absence of infected pairs in the control group. To interpret some results of the published studies in Table 1which concluded difference in the transmissibility between two viruses, one sample interpretation may better be adopted for each virus against $R_{0,null}=0$ (rather than comparing two sample groups). Over-interpretation of results without significant difference should be avoided.

It should be noted that demonstrating similar transmissibility between two groups is even more difficult in this context for two reasons. First, the sample size (i.e. the number of pairs) is very limited, and thus, it is too frequent that we do not observe any significant difference between two sample groups (Table 3). Second, demonstrating similarity must be distinguished from showing the absence of a significant difference, especially when the sample sizes are very limited. Demonstrating similarity in transmissibility would likely require much larger sample sizes, as seen in noninferiority and equivalence randomized trials [32]. Similarly, demonstrating the absence of substantial transmissibility in a single sample group is also a difficult task. As was shown in Table 2, showing $R_0<1$ cannot be achieved by $n=3$, 4 or 5.

Two limitations should be noted. First, every transmission experiment study examines not only the frequency of successful transmissions but also other factors including mortality, weight loss, patterns in virus shedding, clinical signs and symptoms, behavioral changes and so on. Thus, although the quantification of transmissibility should strictly adhere to the frequency of transmission events, the interpretation of adaptation, pathogenicity and infectiousness (e.g. in the sense of virus replications in infected hosts) should be judged from multiple results. Moreover, study objectives of transmission experiment may not necessarily be to demonstrate differential transmissibility (e.g. may only be to prove that pre-symptomatic transmission can occur [33]). In this regard, the present study has focused only on a single aspect of experimental findings. Second, we adopted an independence assumption between pairs which may not be strictly the case in all published studies. If the transmission studies handled infectious virus with substantially high aerosol transmission potential, this assumption is violated. However, explicitly addressing this point cannot be achieved by employing a simple general stochastic epidemic model, and rather, would require much more complex modeling analysis. Resolving second problem (of dependence between pairs) by mathematical modeling are our forthcoming future studies.

It is important that the researchers can calculate the most appropriate sample size (with or without a reference sample group) depending on study objectives. Sample size rationale should be formulated in the future including other types of transmission experiment design, such as exposing multiple animals to multiple infectious animals, or exposing two animals in two different manners (e.g. with and without direct contact). As the translation of the proportion of infected pairs into R_0 can permit object-oriented experimental design, the objectives, design and findings of transmission experiment studies have to be reviewed and the corresponding sample size rationale should be discussed. The present study could be regarded as a starting point to extensively consider hypothesis testing of animal transmission experiment results using epidemiologically well-defined parameters.

AUTHOR CONTRIBUTIONS

Conceived and designed the experiments: HN. Performed the experiments: HN. Analyzed the data: HN. Contributed reagents/ materials/analysis tools: HN HLY BJC. Wrote the paper: HN HLY BJC.

REFERENCES

1. Diekmann O, Heesterbeek JAP (2000) Mathematical Epidemiology of Infectious Diseases: Model Building, Analysis and Interpretation. New York: Wiley.

2. De Jong MC, Kimman TG (1994) Experimental quantification of vaccine-induced reduction in virus transmission. Vaccine 128: 761–766.

3. Imai M, Watanabe T, Hatta M, Das SC, Ozawa M, et al. (2012) Experimental adaptation of an influenza H5 HA confers respiratory droplet transmission to a reassortant H5 HA/H1N1 virus in ferrets. Nature. 486: 420–428.

4. Russell CA, Fonville JM, Brown AE, Burke DF, Smith DL, et al. (2012) The potential for respiratory droplet-transmissible A/H5N1 influenza virus to evolve in a mammalian host. Science. 336: 1541–1547.

5. Herfst S, Schrauwen EJ, Linster M, Chutinimitkul S, de Wit E, et al. (2012) Airborne transmission of influenza A/H5N1 virus between ferrets. Science. 336: 1534–1541.

6. Killingley B, Enstone J, Booy R, Hayward A, Oxford J, et al. (2011) Potential role of human challenge studies for investigation of influenza transmission. Lancet Infect Dis. 11: 879–886.

7. Lee VJ, Yap J, Cook AR, Chen MI, Tay JK, et al. (2010) Effectiveness of public health measures in mitigating pandemic influenza spread: a prospective sero-epidemiological cohort study. J Infect Dis. 202: 1319–1326.

8. Van Hoeven N, Pappas C, Belser JA, Maines TR, Zeng H, et al. (2009) Human HA and polymerase subunit PB2 proteins confer transmission of an avian influenza virus through the air. Proc Natl Acad Sci U S A. 106: 3366–3371. doi: 10.1073/pnas.0813172106

9. Tellier R (2009) Aerosol transmission of influenza A virus: a review of new studies. J R Soc Interface. 6: S783–S790.

10. Belser JA, Katz JM, Tumpey TM (2011) The ferret as a model organism to study influenza A virus infection. Dis Model Mech. 4: 575–579.

11. Maines TR, Chen LM, Matsuoka Y, Chen H, Rowe T, et al. (2006) Lack of transmission of H5N1 avian-human reassortant influenza viruses in a ferret model. Proc Natl Acad Sci U S A. 103: 12121–12126.

12. Tumpey TM, Maines TR, Van Hoeven N, Glaser L, Solórzano A, et al. (2007) A two-amino acid change in the hemagglutinin of the 1918 influenza virus abolishes transmission. Science. 315: 655–659.

13. Munster VJ, de Wit E, van den Brand JM, Herfst S, Schrauwen EJ, et al. (2009) Pathogenesis and transmission of swine-origin 2009 A(H1N1) influenza virus in ferrets. Science. 325: 481–483.

14. Maines TR, Jayaraman A, Belser JA, Wadford DA, Pappas C, et al. (2009) Transmission and pathogenesis of swine-origin 2009 A(H1N1) influenza viruses in ferrets and mice. Science. 325: 484–487.

15. Pappas C, Viswanathan K, Chandrasekaran A, Raman R, Katz JM, et al. (2010) Receptor specificity and transmission of H2N2 subtype viruses isolated from the pandemic of 1957. PLoS One. 5: e11158.

16. Kiso M, Shinya K, Shimojima M, Takano R, Takahashi K, et al. (2010) Characterization of oseltamivir-resistant 2009 H1N1 pandemic influenza A viruses. PloS Pathog. 6: e1001079.

17. van Doremalen N, Shelton H, Roberts KL, Jones IM, Pickles RJ, et al. (2011) A single amino acid in the HA of pH1N1 2009 influenza

virus affects cell tropism in human airway epithelium, but not transmission in ferrets. PLoS One. 6: e25755. doi: 10.1371/journal.pone.0025755

18. Koster F, Gouveia K, Zhou Y, Lowery K, Russell R, et al. (2012) Exhaled aerosol transmission of pandemic and seasonal H1N1 influenza viruses in the ferret. PLoS One. 7: e33118.

19. Pearce MB, Jayaraman A, Pappas C, Belser JA, Zeng H, et al. (2012) Pathogenesis and transmission of swine origin A(H3N2)v influenza viruses in ferrets. Proc Natl Acad Sci U S A. 109: 3944–3949.

20. Billard L, Zhao Z (1993) The stochastic general epidemic revisited and a generalization. IMA J Math Appl Med Biol 10: 67–75.

21. Bartlett MS (1949) Some evolutionary stochastic processes. J R Stat Soc Ser B 11: 211–229.

22. Velthuis AG, Bouma A, Katsma WE, Nodelijk G, De Jong MC (2007) Design and analysis of small-scale transmission experiments with animals. Epidemiol Infect. 135: 202–217.

23. Nishiura H, Klinkenberg D, Roberts M, Heesterbeek JAP (2009) Early epidemiological assessment of the virulence of emerging infectious diseases: A case study of an influenza pandemic. PLoS ONE 4: e6852.

24. Velthuis AG, de Jong MC, de Bree J, Nodelijk G, van Boven M (2002) Quantification of transmission in one-to-one experiments. Epidemiol Infect 128: 193–204.

25. Thomas RG, Conlon M (1992) Sample size determination based on Fisher's Exact Test for use in 2×2 comparative trials with low event rates. Control Clin Trials. 13: 134–147.

26. Agresti A, Coull BA (1998) Approximate is better than 'exact' for interval estimation of binomial proportions. Am Stat 52: 119–126.

27. Krishnamoorthy K, Peng J (2007) Some properties of the exact and score methods for binomial proportion and sample size calculation. Communications in Statistics –Simulation and Computation 36: 1171–1186.

28. Nishiura H (2012) Early detection of nosocomial outbreaks caused by rare pathogens: A case study employing score prediction interval. Osong Public Health and Research Perspectives 3: 121–127.

29. Nishiura H, Yan P, Sleeman CK, Mode CJ (2012) Estimating the transmission potential of supercritical processes based on the final size distribution of minor outbreaks. J Theor Biol 294: 48–55.

30. Klick B, Nishiura H, Ng S, Fang VJ, Leung GM, et al. (2011) Transmissibility of seasonal and pandemic influenza in a cohort of households in Hong Kong in 2009. Epidemiology 22: 793–796.

31. Nishiura H, Chowell G, Castillo-Chavez C (2011) Did modeling overestimate the transmission potential of pandemic (H1N1-2009)? Sample size estimation for post-epidemic seroepidemiological studies. PLoS ONE 6: e17908.

32. Piaggio G, Elbourne DR, Altman DG, Pocock SJ, Evans SJ, et al. (2006) Reporting of noninferiority and equivalence randomized trials: an extension of the CONSORT statement. JAMA. 295: 1152–1160.

33. Roberts KL, Shelton H, Stilwell P, Barclay WS (2012) Transmission of a 2009 H1N1 pandemic influenza virus occurs before fever is detected, in the ferret model. PLoS One. 7: e43303.

CHAPTER 12

Prevalence of Coxiella burnetii in environmental samples collected from cattle farms in Eastern and Central Poland (2011–2012)

Agata Bielawska-Dro´ zd a, *, Piotr Cies´lik a , Tomasz Mirski a , Jerzy Gaweł a , Aleksander Michalski a , Marcin Niemcewicz a , Michał Bartoszcze a , Dorota Z˙ akowska a , Krzysztof Lasocki a , Jo´ zef Knap b , Janusz Kocik a

a Biological Threats Identification and Countermeasure Centre of Military Institute of Hygiene and Epidemiology, Pulawy, Poland

b Institute of Social Medicine, Department of Epidemiology, Medical University of Warsaw, Warsaw, Poland

ABSTRACT

Coxiella burnetii is the etiologic agent of Q fever. It may occur as two different morphological forms, a large cell variant (LCV) and a small cell variant (SCV). The SCV is characterized by unique resistance to physical and chemical factors and may survive in the environment for many months.

The objective of this study was to examine environmental samples for the presence of *C. burnetii* using real-time PCR in areas where Q fever was previously reported and in randomly selected animal farms where Q fever was not reported. The samples were collected in the following provinces in Poland: Lublin, Subcarpathian and Masovian. Monitoring was performed with real-time PCR and serological methods. Of the 727 environmental samples, 33 (4.54%) contained the multi-copy insertion sequence *IS1111*, which is specific for *C. burnetii.* Subsequently, the presence of *C. burnetii* antibodies was determined using serological tests in selected herds in which positive genetic results were obtained. Serological analyses of 169 serum samples using CFT and ELISA were performed on Polish black-and-white Holstein-Friesian cows and one cow imported from Denmark. Using the CFT method, 11 samples were positive for phase I antibodies and six were positive for phase II antibodies. Moreover, in two cases, the presence of antibodies specific for both phase I and phase II antigens of *C. burnetii* was detected. However, of the 169 examined serum samples, 20 were positive by ELISA test, of which six were also positive by CFT. Additionally, multi spacer typing (MST) of isolated *C. burnetii* strains was performed. The MST results identified two new genotypes in Poland, ST3 and ST6. The results indicate that continued research regarding spread of this pathogen within a country is necessary.

KEYWORDS

Coxiella burnetii; Zoonosis; Environmental samples; *IS1111*; Real-time PCR; Multi spacer typing

1. INTRODUCTION

Coxiella burnetii is a small, polymorphic Gram-negative bacterium from 0.3 to 0.5 microns in size, belonging to the order Legionellales, family Coxiellaceae, and genus *Coxiella* (Garrity et al., 2005). Important sources of human infection of *C. burnetii* are small ruminants, such as sheep and goats, dogs, cats, guinea pigs and gerbils (Bielawska-Drózd et al., 2013). According to recent studies, *C. burnetii* shows great phylogenetic similarity to bacteria from the order Legionellales. *C. burnetii* occurs in two different morphological forms, LCV (large cell variant) and SCV (small cell variant) (Mallavia and Minnick, 1995). *C. burnetii* SCV is characterized by unique resistance to physical factors, such as ultraviolet radiation, and chemical factors.

In Poland, several Q fever outbreaks were reported during last 60 years in Owczary, formerly Rzeszów Province in 1956, Lesser Poland and Silesia Provinces in 2005, and Subcarpathian and Lublin Province in 2008. The largest Q fever outbreak occurred in 1982 in Ulhówek, formerly Zamość Province, and resulted in 220 dairy cow cases and 1300 human cases, which was the largest Q fever outbreak in the human population of Poland (Bielawska-Drózd et al., 2013). According to Polish legislation, Q fever is not a disease that is mandatory for control and prevention (isap.sejm.gov.pl, Dz.U. z 2004, r. Nr 69, poz. 625). Despite the availability of Coxevac, vaccinations against Q fever in livestock are not routinely implemented.

For detailed molecular characterization of the agent, phylogenetic method such as MST (multi spacer typing) was used (Glazunova et al., 2005 and Li et al., 2009).

The objective of this study was to examine environmental samples for the presence of *C. burnetii* using real-time PCR and genetic characterization of isolated strains using the MST method.

Subsequently, the presence of *C. burnetii* antibodies was investigated using serological tests (CFT and ELISA) in selected herds in which positive genetic results were obtained.

2. MATERIALS AND METHODS

2.1. Environmental samples

Environmental samples were collected from eight livestock farms where Q fever was previously reported by the Local Veterinarians Inspectorate, and additional samples were obtained from 17 randomly selected control farms. These farms were located in the Masovian, Lublin and Subcarpathian Provinces (Fig. 1).

All 727 environmental samples were collected from the cowshed and represented contaminated manure, urine and amniotic fluid, and swab samples collected from walls, feeders, floors and other areas in direct contact with the farm animals. The swab samples were placed in 1 ml sterile saline solution. The samples were collected during the spring and summer of 2011–2012.

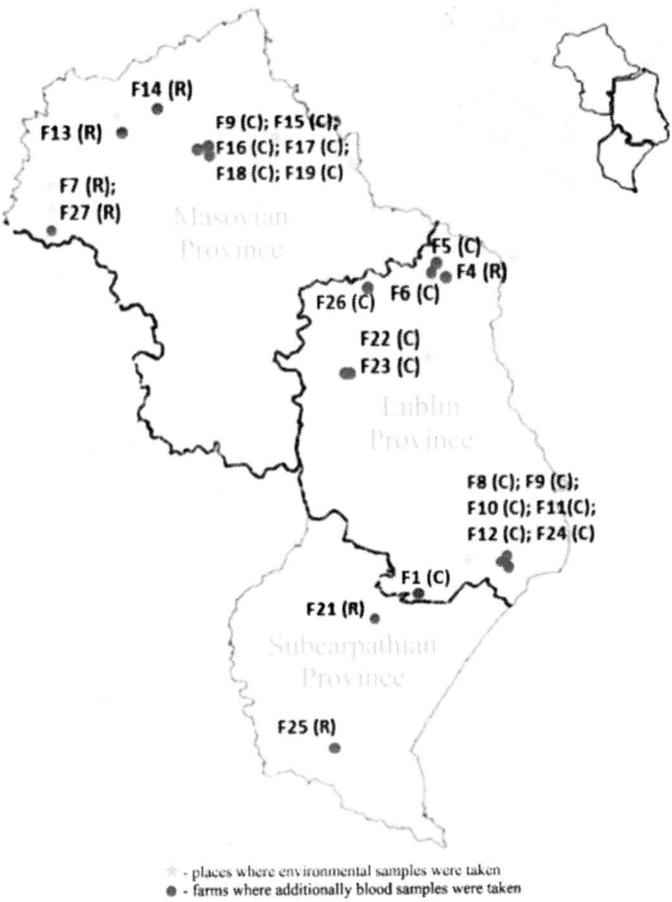

Figure 1. Places of origin of the environmental and serum samples. Legend: F1 (C), Biłgoraj District; Tranogród Commune, F4 (R), Bielsko, Podlaski District; Witulin Commune, F7, F27 (R), Gostynin District; Pacyna Commune, Luszyn Village; F13 (R), Ciechanowski District; Glinojeck Commune, Glinojeck Village; F14 (R), Ciechanowski District; Ciechanów Commune, Kargoszyn Village; F21 (R), Leżajsk District; Dębno Village; F25 (R), Brzozów District; Haczów Commune; F5 (C), Bialski District; Worgule Commune; F6 (C), Bialski District; Łukowce Commune; F8 (C), Tomaszów Lubelski District; Rachanie Commune, Siemnice Village; F9 (C), F10 (C), Tomaszów Lubelski District; Rachanie Commune, Michałów Village (2 farms); F11 (C), F12 (C), Tomaszów Lubelski District; Rachanie Commune, Wożuczyn Village (2 farms); F24 (C), Tomaszów Lubelski District; Rachanie Commune, Podlodów Village; F22, F23 (C), Puławski District; Baranów Commune, Drążgów Village (2 farms); F26 (C), Łuków District; Trzebieszów Commune, Wierzejki Village; F15, F16 (C), Pułtusk District; Brodowo-Bąboły (2 farms); F17 (C), F18 (C), F19 (C), F9 (C), Pułtusk District; Łachoń, Skarżyce, Pawłowo Village (2 farms).

2.2. Vaginal swabs

Because of an unexpected abortion in the F26 farm (Wierzejki Village, Lublin Province), 30 vaginal swab samples were collected from cows. The material was collected using cotton swabs.

2.3. Blood samples

Blood samples were collected from the eight herds of cattle in which the presence of *C. burnetii* DNA in the local environment was previously detected by real-time PCR, excluding farm F26. The samples were primarily collected from Polish black-and-white Holstein–Fresian cows, with an average age of 6 years, with 75% of individuals pregnant, and from one cow imported from Denmark. In summary, for the laboratory tests, 169 bovine blood samples were collected, from which sera were obtained and then examined for the presence of anti-*C. burnetii* antibodies using CFT and ELISA. CFT was performed to detect the presence of antibodies against phase I and phase II antigens according to manufacturer's recommendations (Institute Virion/Serion GmbH, Germany) and the Institute of Biotechnology, Sera and Vaccines (IBSS Biomed, Poland). Sera were diluted in a range of 1:5–1:80. The samples were considered positive when partial hemolysis (++) at a dilution of 1:10 was observed. The study and interpretation of the ELISA results (IDEXX Laboratories Inc., USA) was performed according to the manufacturer's instructions.

For blood sampling, local veterinarians obtained consent from farm owners. Ethical approval was not required for the collection of blood samples.

2.4. Environmental samples

For DNA isolation, the samples were placed in saline solution in a volume of not more than 2 ml. The samples were then centrifuged at 800 $\times g$ for 10 min (decanting) to remove any large organic residues, and the supernatant was transferred to a new sterile tube. A volume of 1 ml suspension was placed into a test tube and subjected "heat shock" in saline solution containing 2 mM adenosine and 10 mM L-alanine (62.5 °C for 15 min.). Next, 4 ml of liquid MEM (Minimum Essential

Medium; Sigma-Aldrich, USA) was added to the sample, and the suspension was subjected to 3-fold filtering through a 0.45 mm filter. After filtration, the obtained material was centrifuged for 10 min at 25,000 × *g* to pellet the cells. The supernatant was discarded and the pellet was resuspended in 2 ml MEM. The suspension was used to infect the cell line (1 ml), whereas 700 µl was used for the isolation of genetic material using the Genomic Mini Kit (A&A Biotechnology, Poland) according to the manufacturer's instructions.

2.5. Vaginal swabs

The isolation of genetic material was performed similar to the environmental swab samples from the surface, but without pre-spinning. The tested samples were transferred to new 1.5-ml tubes, centrifuged at 20,000 × *g*, and the pellet was resuspended in 100 µl of Tris-HCl. DNA isolation was performed as previously described.

2.6. Real-time PCR

Genetic analysis was performed using real-time PCR to detect the presence of the insertion sequence *IS1111* (transposase gene, GenBank: M80806) (Klee et al., 2006) on a LightCycler 2.0 instrument (Roche, Germany) using the LightCycler TaqMan Master Kit (Roche, Germany) according to manufacturer's recommendations with final concentrations of 0.5 µM for the primers and 0.15 µM for the probe.

As a positive control, the DNA extracted from the *C. burnetii* strain Nine Mile phase I was used, and as a reaction control, the mixture without template DNA was used. Real-time PCR was performed for 40 cycles, and the test samples for which the C_T (cycle threshold) was lower than 36 were considered positive.

2.7. BGM cell line culturing

Culturing of the BGM cell line (African green monkey kidney, HPA Culture Collection, UK) was performed at 37 °C in a 5% CO_2

atmosphere with MEM containing 2 mM glutamine (Sigma Aldrich, USA), 1% non-essential amino acids (NEAA) (Sigma Aldrich, USA), and 10% serum (fetal bovine serum; FBS) (Sigma Aldrich, USA), 100 μg/ml streptomycin (Sigma Aldrich, USA) and 100 U/ml penicillin (Sigma Aldrich, USA).

2.8. *C. burnetii* culturing in BGM cell line

Culturing of *C. burnetii* was performed in the BGM cell line in 25 cm^3 culture flasks incubated under aerobic conditions in a 5% CO_2 atmosphere with MEM medium containing 10% FBS, 1 mM L-glutamine and 1% NEAA. Culturing was performed for up to 14 days. No antibiotics were added to the culture medium.

2.9. *C. burnetii* recovery from the cell line

To recover *C. burnetii* from the cell line, the method using digitonin described by Cockrell et al. (2008) was used. The obtained lysate was centrifuged at 900 × *g* for 5 min at 4 °C. The supernatant containing the released *C. burnetii* cells were transferred to a new 50-ml tube and centrifuged at 31,000 × *g* for 15 min at 4 °C. The supernatant was discarded, and the pellet was resuspended in 2 ml MEM with 10% FBS (Sigma-Aldrich, USA). All samples were stored at −80 °C until further analysis.

2.10. Isolation of *C. burnetii* DNA after digitonin treatment

For DNA extraction, 1 ml of lysate was removed and centrifuged at 15,000 × *g* for 5 min for cell collection. The supernatant was discarded and the pellet was suspended in 100 μl Tris–HCl buffer. The subsequent steps were performed according to Genomic Mini kit manufacturers' instructions.

2.11. Multi Spacer Typing

Six of 10 (Cox18, Cox20, Cox37, Cox51, Cox57, Cox61) proposed intergenic spacers previously described by Glazunova et al. (2005) were selected for MST analysis of the recovered isolates.

2.12. Sequence analysis

The obtained sequences were compared with the allelic profiles for the selected intergenic sequences available in the IFR 48 using the BLAST algorithm at the NCBI website (http://blast.ncbi.nlm.nih.gov/Blast.cgi). Phylogenetic analysis was performed using the MEGA5 software (Tamura et al., 2011).

2.13. Ethics statement

Animal sampling, including serum and vaginal sample collection, was performed under control of the Regional and Local Veterinary Inspectorates (Masovian, Subcarpathian and Lublin) after receiving the consent of the farmer. According to Polish Ethical Committee on Animal Testing, this type of study does not require consent or ethical approval.

3. RESULTS

Of the 8 livestock farms, F1, F4, F7, F13, F14, F21, F25, and F27, in which Q fever was detected in animals previously, 6 farms, including F4, F7, F13, F14, F21, and F27, had the genetic markers for *C. burnetii* in the environmental samples (Fig. 1). Of the 409 tested environmental samples from the retrospective farms, 26 (6.36%) had the genetic marker *IS1111*.

For 17 randomly selected control farms from the Provinces in which Q fever has never been reported by the Local Veterinary Inspectorate, 3 farms, F16, F19, and F23, had genetic marker sequences for *C. burnetii* (Fig. 1). From 318 environmental samples derived from these farms, 7 (2.2%) had *IS1111*.

Of the 727 environmental samples collected from the 25 livestock farms, in 33 (4.54%) we found the *IS1111* gene sequence, which is specific for *C. burnetii*.

Additionally, of the 30 vaginal swabs collected from the cows from the F26 farm in Lublin Province (Fig. 1), one swab (WV 21, C_T 26) from a pregnant cow showed the presence of*IS1111*.

As a result of the serological studies of the four retrospective outbreaks, in two farms, F4 and F13 (Fig. 1), we showed the presence of phase I antibodies and phase II antibodies. Serological analysis was performed for randomly selected individuals (official data regarding specific individuals which were tested in 2010 (Witulin, F13) were not available at this time). Furthermore, of the four randomly selected farms, two (F19, F26– Fig. 1) had *C. burnetii*-specific antibodies in animals, primarily for phase I antibodies. In total, we tested 169 serum samples, among which, 11 were positive for phase I antibodies and 6 positive for phase II antibodies using CFT. Moreover, in two cases, we showed the presence of antibodies specific for both phase I and phase II antigens of *C. burnetii*. From 169 samples, 20 were positive by ELISA, of which 6 were also positive by CFT.

The samples positive for the *IS1111* gene sequence were used to infect the BGM cells *in vitro*. A positive fluorescence signal for the isolates obtained from the BGM cell line was obtained for the samples derived from the new outbreaks in the F19 farm in Masovian Province (sample 124 WPO; C_T 23) and in the F26 farm in Lublin Province (sample WV 21; C_T 26). For the *IS1111*+ swabs derived from the retrospectively examined farms, we obtained 4 isolates, 339 WPO with C_T 28 (F21), 635 WPO with C_T 31 (F27), 640 WPO with C_T 29 (F27) and 86 WPO with C_T 32 (F13) by *in vitro* culture of the BGM cell line.

For the isolates derived from cow vaginal swab WV 21 (F26) and an environmental area swab 124 WPO (F19), an MST analysis using 6 target sequences was performed. The obtained sequences were compared with those deposited on IFR 48. The ST profiles for the examined strains and reference strains and their genetic relatedness are presented inFig. 2. The results identified their MST genotype. The WV 21 sample was the ST6 genotype, whereas the 124 WPO sample was the ST3 genotype.

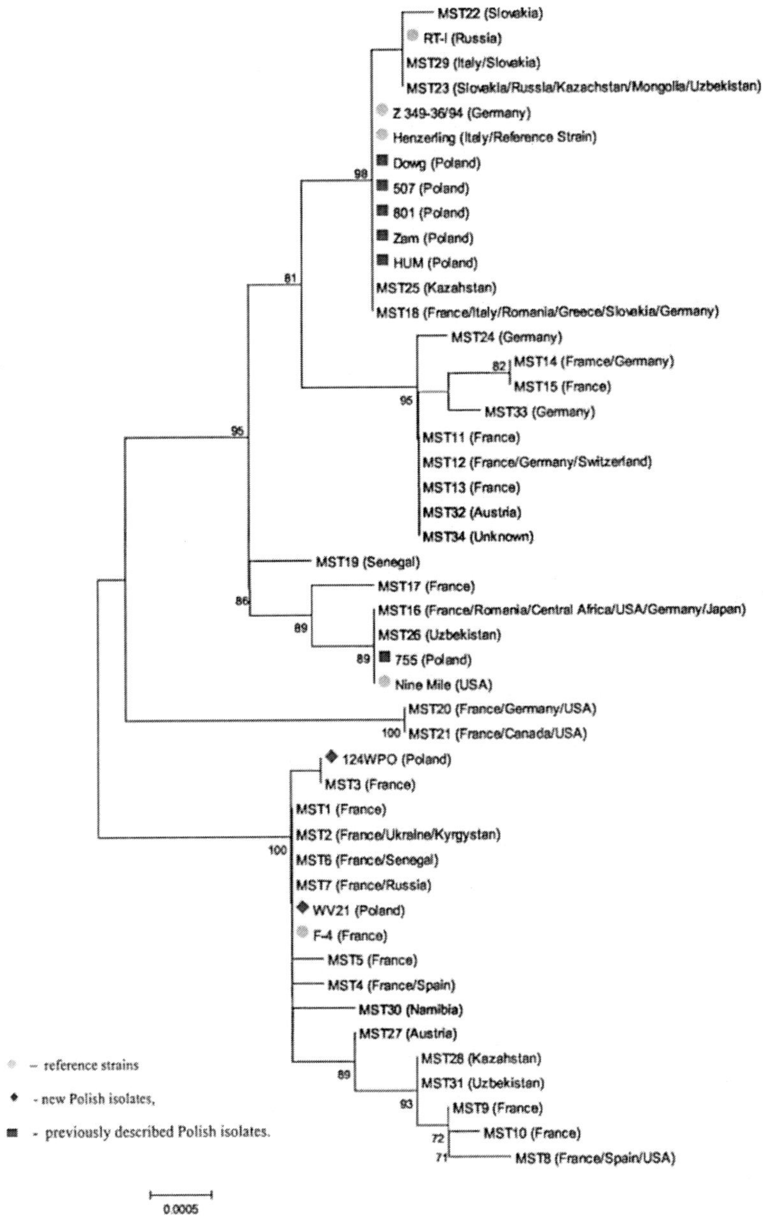

Origin of other genotypes were described according to Mediannikov et al. (2010)

Figure 2. Genetic relatedness of Polish *C. burnetii* isolates, WV 21 and 124 WPO, compared to isolates described by Mediannikov et al. (2010) and Chmielewski et al. (2009). Legend: The tree was drawn using the Neighbor-Joining method with a maximum composite likelihood model (unrooted) with 500 replications. The sequences from 6 spacers were concatenated. Bootstrap values >70 are shown. The scale bar shows the number of substitutions per site. The tree was drawn with the MEGA5 software.

4. DISCUSSION

In this study, the presence of Q fever agent was detected in the environment in the majority of retrospective farms and only several of the randomly selected control farms. Real-time PCR was successfully used with oligonucleotides specific for the transposase gene (*IS1111*) to determine the presence of *C. burnetii* in the environmental samples.

Environmental screening for the presence of *C. burnetii*, was performed in six distant regions in the United States from 2006 to 2008 (Kersh et al., 2010). Positive results were detected in 23.8% of 1600 analyzed samples, and depending on the region, varied from 6% to 44.6%. Research was conducted in livestock farms and public places, such as high schools, playgrounds, markets, banks, and post offices. Furthermore, similar results were obtained by Tozer et al. (2013) who examined environmental sources, such as soil and swab samples, of *C. burnetii* infection for both rural and urban populations in Queensland, Australia. The positive results ranged from 2.0 to 6.9% (Tozer et al., 2013). Although the detection of *C. burnetii* in livestock farms of goats and cattle was not unexpected (de Bruin et al., 2012, Hogerwerf et al., 2012 and Kersh et al., 2013), the presence of the agent in public areas was surprising and indicated a risk for humans (Kersh et al., 2010). Our results showing that 2.2% of the samples were positive in the randomly selected control farms, confirm that the natural environment is not free of *C. burnetii* and Q fever risk, even though no cases are currently registered for a specific area.

It was observed that there are more contaminated farms where Q fever cases were reported previously than in the randomly selected farms. This may be a result of not only the inappropriate eradication disinfection process or pathogen resistance but also natural reservoirs. Therefore, in areas that were determined as Q-fever free, should occasionally be reevaluated for agent persistence in both biotic and abiotic components of the environment.

Seroprevalence studies of the animals revealed seropositive individuals in both the retrospective and randomly selected farms. Unfortunately, we could not examine all herds because of the lack of owner consent for blood sampling. Special attention was paid to farm F13, where seroprevalence studies in 2010 detected anti-*C. burnetii* antibodies in one case, not confirmed in the official analysis. Molecular analysis of the environment revealed a significant number (23.3%) of *C. burnetii* positive samples. A total of 49 animal sera were tested for specific antibodies against *C. burnetii* using CFT and ELISA. Only 4 (8.3%) sera samples were positive for antibodies against phase II antigens when tested with CFT, whereas ELISA revealed an additional 13 (27.1%) sera samples positive for specific antibodies against *C. burnetii*. The lower detection limit of the complement fixation test compared to the ELISA is consistent with the literature (Horigan et al., 2011,Natale et al., 2012, Rousset et al., 2009 and Szymańska-Czerwińska et al., 2013). The lack of positive correlation for the CFT and ELISA results was confirmed by statistical analysis. However, CFT is the method of choice for official diagnosis by the OIE (www.oie.int). Some infected animals, which are seropositive by CFT, do not shed pathogen, whereas other animals are infected but seronegative by CFT and can shed the agent into the environment (Sidi-Boumedine et al., 2010). Serological methods are useful for Q fever diagnosis, but cannot be used to identify *C. burnetii* shedding animals (Rousset et al., 2009 and Rodolakis, 2006).

The results of the randomly selected F26 farm will be discussed in greater detail. Cattle abortion at the eighth month of gestation was observed in a cow of the Polish black-and-white breed. The fetus and placenta were diagnosed using molecular methods on May 2012, and the placenta was positive for the genetic sequence of *C. burnetii*. After this incident, the decontamination process was implemented by owner in the cowshed, which may have influenced the negative real-time PCR results for presence of *C. burnetii* in this farm. Afterwards, 30 vaginal swabs were tested using real-time PCR. The only swab taken from a pregnant cow (WV 21 sample), which was standing next to the above described aborting cow, was positive for the genetic sequence of *C. burnetii*, but negative by CFT and ELISA. This phenomenon may be an effect of an early phase infection related to the time-frame of seroconversion (3–4 weeks). Additionally, six other animals in this herd were positive, five for antibodies against phase I and one for phase II antibodies.

Seroprevalence studies in the new F19 farm using CFT revealed 2 positive cases for phase I antibodies out six samples tested. The presence of *C. burnetii* genetic material was also detected in 3 of 23 environmental samples. Environmental isolate (124 WPO) was successfully recovered using *in vitro* cell culture.

Farmers and veterinarians are at risk of Q-fever infection. It is hypothesized that seroprevalence studies of people will provide significant insight; however, this was beyond the scope of our study. Independent serological studies of the human population in urban areas revealed a 3.1% seropositive rate for antibodies against *C. burnetii*. However, based on the level of *C. burnetii* in the environment (44%), the actual seroprevalence may be higher (Kersh et al., 2010).

MST, which utilizes the variability of intergenic sequences, may be successfully used for high resolution typing of bacterial strains and the assessment of the relationship between the genetic signature and geographical distribution. Polish *C. burnetii* isolates, previously recovered, were genotyped as ST16 (1 isolate) and ST18 (5 isolates) (Chmielewski et al., 2009). In the current study, genotypes ST3 and ST6 were identified (Fig. 2). Moreover, genotype ST6 was also detected in ticks and fecal specimens from Senegal, and in a specimen from an endocarditis patient from France. ST3 was also found in France (Mediannikov et al., 2010). Detection of these two genotypes suggests the genetic variability of *C. burnetii* in Poland, but further research is necessary to understand the genetic diversity and identify methods to control possible Q fever outbreaks. Interestingly, genotypes of WV 21 (ST6) and 124 WPO (ST3) were clustered with genotypes isolated from other countries, including France, Spain, Austria, Ukraine, Russia, Kyrgyzstan, Uzbekistan, Kazakhstan, Namibia, Senegal and the USA. It is postulated that genotypes ST3 and ST6 have a worldwide distribution (Mediannikov et al., 2010).

This study shows the presence of *C. burnetii* in livestock farms in which the disease was detected previously and in farms where Q fever has never been detected in Eastern and Central Poland. This may play a role in the increasing risk of potential infections for human populations.

Based on the results of this study, ELISA is a more sensitive method than CFT for *C. burnetii* antibody detection.

Genetic analysis of the environmental isolates using MST revealed two genetic profiles, ST3 and ST6, which were previously described in other parts of Europe and the world, but for the first time in Poland.

CONFLICTS OF INTEREST

Authors declare that they have no conflicts of interest.

ACKNOWLEDGMENTS

We acknowledge support from the Subcarpathian Regional Veterinary Inspectorate in Krosno (Mirosław Weltz, PhD, DVM) and Lublin Regional Veterinary Inspectorate (Jerzy Zarzeczny, DVM). Reference strains of *C. burnetii* (F4, Nine Mile, RT-1 and Z349-36/94) were kindly provided by Dr. Dimitrios Frangoulidis and Dr. Holger Scholz from the Microbiology Institute of German Armed Forces (Munich, Germany). We would like to acknowledge Anna Roszkowiak who performed the sample collection and preparation and Monika Szymańska-Czerwińska and Krzysztof Niemczuk for serological analysis. We thank laboratory technicians Barbara Ciecierska and Dorota Sikorska for technical assistance. The study was funded by the Ministry of Science and Higher Education of the Republic of Poland (Grant No. N N308 562540).

REFERENCES

1. Bielawska-Dro´ zd, A., Cies´lik, P., Mirski, T., Bartoszcze, M., Knap, J.P., Gaweł, J., Z˙ akowska, D., 2013. Q fever - selected issues. Ann. Agric. Environ. Med. 20, 222–232.

2. Chmielewski, M., Sidi-Boumedine, K., Duquesne, V., Podsiadły, E., Thiery, R., Tylewska-Wierzbanowska, S., 2009. Molecular epidemiology of Q fever in Poland. Pol. J. Microbiol. 58, 9–13.

3. Cockrell, D.C., Beare, P.A., Fischer, E.R., Howe, D., Heinzen, R.A., 2008. A method for purifying obligate intracellular Coxiella burnetii that employs digitonin lysis of host cells. J. Microbiol. Methods 72, 321–325.

4. de Bruin, A., van der Plaats, R.Q.J., de Heer, L., Paauwe, R., Schimmer, B., Vellema, P., van Rotterdam, B.J., van Duynhoven, Y.T.H.P., 2012. Detection of Coxella burnetti DNA on small-ruminant farms during a Q fever outbreak in the Netherlands. Appl. Environ. Microbiol. 78, 1652–1657.

5. Garrity, G., Brenner, D.J., Krieg, N.R., Staley, J.R., 2005. Bergey's Manual of Systematic Bacteriology. Springer-Verlag.

6. Glazunova, O., Roux, V., Freylikman, O., Sekeyova, Z., Fourous, G., Tyczka, J., Tokarevich, N., Kovacova, E., Marrie, T.J., Raoult, D.,

2005. Coxiella burnetii genotyping. Emerg. Infect. Dis. 11, 1211–1217.

7. Hogerwerf, L., Borle´e, F., Still, K., Heederik, D., van Rotterdam, B., de Bruin, A., Nielen, M., Wouters, I.M., 2012. Detection of Coxiella burnetti DNA in inhalabel airborne dust samples from goat farms after mandatory culling. Appl. Environ. Microbiol. 78, 5410–5412.

8. Horigan, M.W., Bell, M.M., Pollard, T.R., Sayers, A.R., Pritchard, G.C., 2011. Q fever diagnosis in domestic ruminants: comparison between complement fixation and commercial enzyme-linked immunosorbent assays. J. Vet. Diagn. Invest. 23, 924–931.

9. isap.sejm.gov.pl, Dz.U. z 2004, r. Nr 69, poz. 625, Ustawa z dnia 11 marca 2004 r o ochronie zdrowia zwierza˛ t oraz zwalczaniu choro´b zakaz´ - nych zwierza˛ t.

10. Kersh, G.J., Fitzpatrick, K.A., Self, J.S., Priestley, R.A., Kelly, A.J., Lash, R.R., Marsden-Hung, N., Nett, R.J., Bjork, A., Massung, R.F., Anderson, A.D., 2013. Presence and persistence of Coxiella burnetti in the environments of goat farms associated with a Q fever outbreak. Appl. Environ. Microbiol. 79, 1697–1703.

11. Kersh, G.J., Wolfe, T.M., Fitzpatrick, K.A., Candee, A.J., Oliver, L.D., Patterson, N.E., 2010. Presence of Coxiella burnetii DNA in the environment of the United States, 2006 to 2008. Appl. Environ. Microbiol. 76, 4469–4475.

12. Klee, S.R., Tyczka, J., Ellerbrok, H., Franz, T., Linke, S., Baljer, G., Appel, B., 2006. Highly sensitive real-time PCR for specific detection and quantification of Coxiella burnetii. BMC Microbiol. 6, 2.

13. Li, W., Raoult, D., Fournier, P., 2009. Bacterial strain typing in the genomic era. FEMS Microbiol. 33, 892–916.

14. Mallavia, L.P., Minnick, M.F., 1995. Coxiella burnetii as an etiologic agent. In: Anusz, Z. (Ed.), Q Fever in Humans and Animals. Edited by Olsztyn: ART, pp. 70–84.

15. Mediannikov, O., Fenollar, F., Socolovschi, C., Diatta, G., Bassene, H., Molez, J.F., Sokhna, C., Trape, J.F., Raoult, D., 2010. Coxiella

burnetii in humans and ticks in rural Senegal. PLoS Negl. Trop. Dis. 4, 1–8.

16. Natale, A., Bucci, G., Capello, K., Barberio, A., Tavella, A., Nardelli, S., Marangon, S., Ceglie, L., 2012. Old and new diagnostic approaches for Q fever diagnosis: Correlation among serological (CFT, ELISA) and molecular analyses. Comp. Immunol. Microbiol. Infect. Dis. 25, 375–379.

17. Rodolakis, A., 2006. Q fever, state of art: Epidemiology, diagnosis and prophylaxis. Small Rumin. Res. 62, 121–124.

18. Rousset, E., Berri, M., Durand, B., Dufour, P., Prigent, M., Delcroix, T., Rodolakis, A., 2009. Coxiella burnetii shedding routes and antibody response after outbreaks of Q fever-induced abortion in dairy goat herds. Appl. Environ. Microbiol. 75, 428–433.

19. Sidi-Boumedine, K., Rousset, E., Henning, K., Ziller, M., Niemczuk, K., Roest, H.I.J., Thie´ ry, R., 2010. Development of harmonised schemes for the monitoring and reporting of Q fever in animals in the European UnionIn: Scientific Report submitted to EFSA European Food Safety Authority, Parma. , pp. 1–48.

20. Szyman´ ska-Czerwin´ ska, M., Galin´ ska, E.M., Niemczuk, K., Zase˛pa, M., 2013. Prevalence of Coxiella burnetii infection in foresters in the south-eastern Poland and comparison of diagnostic methods. Ann. Agric. Environ. Med. 20, 699–704.

21. Tamura, K., Peterson, D., Peterson, N., Stecher, G., Nei, M., Kumar, S., 2011. MEGA5: Molecular evolutionary genetics analysis using maximum likelihood. evolutionary distance, and maximum parsimony methods. Mol. Biol. Evol. 28, 2731–2739.

22. Tozer, S.J., Lambert, S.B., Strong, C.L., Field, H.E., Sloots, T.P., Nissen, M.D., 2013. Potential animal and environmental sources of Q fever infection for humans in Queensland. Zoon. Pub. Health,

CHAPTER 13

Transfection of RNA from Organ Samples of Infected Animals Represents a Highly Sensitive Method for Virus Detection and Recovery of Classical Swine Fever Virus

Denise Meyer, Stefanie Schmeiser, Alexander Postel, Paul Becher*

EU and OIE Reference Laboratory for Classical Swine Fever,
Institute of Virology, Department of Infectious Diseases,
University of Veterinary Medicine Hannover, Hannover, Germany

ABSTRACT

Translation and replication of positive stranded RNA viruses are directly initiated in the cellular cytoplasm after uncoating of the viral genome. Accordingly, infectious virus can be generated by transfection of RNA genomes into susceptible cells. In the present study, efficiency of conventional virus isolation after inoculation of cells with infectious sample material was compared to virus recovery after transfection of total RNA derived from organ samples of pigs infected with Classical swine fever virus (CSFV). Compared to the conventional method of virus isolation applied in three different porcine cell lines used in routine diagnosis of CSF, RNA transfection showed a similar efficiency for virus rescue. For two samples, recovery of infectious virus was only possible by RNA transfection, but not by the classical approach of virus isolation.

Therefore, RNA transfection represents a valuable alternative to conventional virus isolation in particular when virus isolation is not possible, sample material is not suitable for virus isolation or when infectious material is not available. To estimate the potential risk of RNA prepared from sample material for infection of pigs, five domestic pigs were oronasally inoculated with RNA that was tested positive for virus rescue after RNA transfection. This exposure did not result in viral infection or clinical disease of the animals. In consequence, shipment of CSFV RNA can be regarded as a safe alternative to transportation of infectious virus and thereby facilitates the exchange of virus isolates among authorized laboratories with appropriate containment facilities.

1. INTRODUCTION

Classical swine fever (CSF) is one of the most important diseases of pigs and wild boar worldwide. It is caused by Classical swine fever virus (CSFV), a pestivirus belonging to the family *Flaviviridae* [1]. CSF outbreaks are associated with enormous economic losses, which are a result of the disease itself as well as of implemented control measures including stamping-out strategy and trade restrictions in the European Union (EU) [2, 3]. To confirm a CSF outbreak in the EU, National Reference Laboratories and the EU Reference Laboratory (EURL) have to follow the guidelines as laid down in the Technical Annex of European Commission Decision 2002/106/EC (EU Diagnostic Manual for CSF) [4, 5]. Currently, virus isolation on susceptible cells as well as detection of viral antigen and virus genome are used for CSF diagnosis. Also in case of previous diagnosis of CSF, it is recommended to perform virus isolation. New CSFV isolates have to be sent to the EURL for detailed characterization and are archived along with available epidemiological and sequence information. The CSF virus collection and the corresponding CSF database became an important tool for supporting the common efforts to eradicate CSF in the EU by providing a platform to exchange isolate-related information and by supporting molecular characterization of new CSFV isolates [6–10]. Generally, virus isolation is labor and time intensive and therefore only applicable for the analysis of small sample numbers. Leucocytes, non-coagulated blood and organ samples (lymph node, kidney, spleen, and tonsil) are preferred sample material [4, 5]. However, virus isolation from organ samples is often challenging due to toxic effects on cell culture. Furthermore, microbial contamination can occur after inoculation of cells with organ suspension despite high concentrations of antibiotics in cell culture medium.

However, for detailed virus characterization including determination of virulence, isolation of infectious virus is indispensable.

Pestiviruses contain a single-stranded RNA genome of positive polarity. The genome encompasses one open reading frame which is translated into one polyprotein. Translation and replication of the virus are directly initiated in the cellular cytoplasm after uncoating of the viral genome [11, 12]. Apart from the classical way of infection, virus can be recovered after transfer of viral genomic RNA into cells by liposome-mediated transfection reagents or electroporation. Such transfection methods are the basis for reverse-genetic approaches which allow the generation of infectious virus from *in vitro* transcribed RNAs derived from full-length cDNA clones. Reverse-genetics has been successfully used for molecular characterization of CSFV and related pestiviruses including detailed analyses of genome organization, translation and biosynthesis of viral proteins as well as studies on RNA recombination, cytopathogenicity, and virulence [13–19]. In addition, RNA transfection has also been applied to rescue infectious foot-and-mouth disease virus (FMDV [20–23]) and CSFV [23] from sample material. However, a direct comparison of conventional virus isolation and RNA transfection for virus rescue using sample material of CSFV infected pigs has not been reported so far. In particular for sample material with high microbial contamination potential in cell culture (e.g. organ samples) RNA transfection might represent a good alternative to conventional virus isolation.

In view of the fact that shipment of samples from animals affected with list A diseases of the World Animal Health Organization (Office International des Epizooties; OIE) requires special biosafety precautions, it has been discussed to combine shipment of non-infectious sample material with recovery of infectious virus upon arrival of samples at authorized laboratories [21, 23]. In this context it has been reported that RNA prepared from infectious sample material using Trizol or phenol-chloroform preparation methods may represent a simplified and less expensive alternative to shipment of infectious sample material [23]. However, classification of RNA samples as biological non-hazardous material referred to studies on inoculation of susceptible cells with purified viral RNA. So far, it has not been experimentally examined whether inoculation of susceptible host species with RNA isolated from CSFV positive samples via the oronasal route can lead to infection and disease.

The major goals of the present study were to analyze if RNA transfection can be used as an additional diagnostic approach for isolation of infectious CSFV and whether RNA derived from infectious material poses a threat for pigs. For this purpose, the detection limits of virus rescue by RNA transfection and conventional virus isolation were compared to each other. Both, organ samples of experimentally infected pigs as well as field samples were tested. Furthermore, oronasal inoculation of pigs with RNA, prepared from whole blood of a highly viremic animal, was performed to determine whether accidental contact of pigs with CSFV RNA poses a threat with respect to biosecurity.

2. MATERIAL AND METHODS

2.1 Samples

Organ samples (spleen, kidney, tonsil, and lymph node) of four weaner pigs, which were experimentally infected with different CSFV isolates were analyzed. One animal was infected with cell culture supernatant of CSFV isolate “Rösrath” (CSF1045), two with cell culture supernatant of CSFV strain “Brescia” (CSF0947) and one with whole blood taken from an animal infected with CSFV strain “Koslov”, respectively. These samples were obtained from the organ sample collection of the EURL for CSF (Hannover, Germany). The respective animal experiment for obtaining the samples was notified by the Specialized Department of Animal Welfare Service of the Lower Saxony State Office for Consumer Protection and Food Safety (LAVES; http://www.state-office-for-consumer-protection-and-food-safety.niedersachsen.de/portal/live.php?navigation_id=26144&article_id=73766); Permit Number: LAVES AZ 08A 538) according to the German animal welfare act (Tierschutzgesetz, 25 Mai 1998, § 8a, Abs. 1 u. 2).

In addition, eight field samples [organ suspension, diluted 1:10 in phosphate buffered saline (PBS)] were tested, which were taken from pigs that were culled during a CSF outbreak in Serbia in November 2010 (sample no.: 154–1 to 154–8). Since these samples were collected in the context of CSF control measures no ethical approval was required. The samples were sent by the Scientific Institute of Veterinary Medicine of Serbia, Virology Department, Belgrade, Serbia to the EURL for CSF (Hannover, Germany) to confirm the CSF outbreak. After CSF diagnosis remaining sample material was used in the present study.

2.2 Cells

Three different cell lines [Porcine kidney (PK15), PK15(A) [24], and swine testis epitheloid (STE)] were used for virus isolation. The cell lines PK15 (CCLV0051) and STE (CCLV0255) were obtained from the Collection of Cell Lines in Veterinary Medicine (CCLV, Friedrich-Loeffler-Institute, Island of Riems, Greifswald, Germany). These cell lines are routinely used for virus isolation at the EURL for CSF. Transfection experiments were performed using swine kidney (SK6) cells according to a previously established protocol [17]. SK6 cells were kindly provided by the Institute of Virology and Immunoprophylaxis, Mittelhäusern, Switzerland [25]. PK15(A), PK15 and STE cells were cultivated in Eagle´s Minimum Essential Medium (EMEM) supplemented with 5% [PK15(A)] or 10% fetal calf serum (PK15 and STE) which had been tested for absence of pestivirus genomes and pestivirus-specific antibodies. After infection the cells were maintained in EMEM supplemented with 10% fetal calf serum. SK6 cells were grown in Dulbecco´s modified Eagle´s medium supplemented with 10% fetal calf serum.

2.3 Isolation of viral RNA, real-time RT-PCR, and nucleotide sequencing

Viral RNA was extracted from organ samples using QIAzol (Qiagen, Germany) combined with the RNeasy Mini Kit (Qiagen, Germany). RNA preparation from leucocytes and swab samples was performed with the QIAamp RNA Blood Mini Kit and the QIAamp Viral RNA Mini Kit (Qiagen, Germany) according to the manufacturers' recommendations. The concentration of isolated RNA was determined photometrically (Nanodrop, Thermo Scientific). Viral RNA was detected by a CSFV-specific quantitative reverse transcriptase polymerase chain reaction (qRT-PCR) as described previously [26]. *In vitro* run-off transcripts of target RNA were used as copy standard for quantification of CSFV-specific RNA copies per µg RNA. Afterwards the genome equivalents (RNA copies per electroporation reaction) were calculated on the basis of the RNA amount used for electroporation (Table 1). For sequence analysis, the genomic region encoding the structural protein E2 was amplified by conventional RT-PCR and subjected to nucleotide sequencing as described previously [9].

Table 1. Conventional virus isolation and virus recovery after RNA electroporation testing organ samples from experimentally infected pigs.

CSFV strain (catalogue number [a])	Sample [animal ID-organ-days post infection]	Virus isolation			qRT-PCR		RNA transfection of SK6 cells		
		PK15	PK15 (A)	STE	C_q value	copies /µg total RNA	Amount of total RNA [µg/ reaction]	Genome equivalents [copies/reaction]	Recovery of infectious virus[b]
Rösrath (CSF1045)	261-spleen-26	+	+	+	23.05	4.2×10^4	8.8	3.7×10^5	1:4,000
	261-kidney-26	+	+	+	24.07	1.9×10^4	10.5	1.9×10^5	1:40
	261-tonsil-26	+	+	+	20.76	1.5×10^5	17.6	2.7×10^6	1:400
	261-lymph node-26	+	+	+	20.27	1.1×10^5	32.9	3.7×10^6	1:4,000
Brescia (CSF0947)	283-spleen-09	+	+	+	21.62	5.8×10^5	2.6	1.5×10^6	1:400
	283-kidney-09	+	+	+	27.10	2.9×10^3	8.1	2.4×10^4	-
	283-tonsil-09	+	+	+	26.68	3.4×10^4	0.9	3.0×10^4	1:4
	283-lymph node-09	+	+	+	25.53	2.3×10^4	3.0	6.9×10^4	-
	284-spleen-07	+	+	+	27.23	9.2×10^3	2.4	2.2×10^4	1:40
	284-kidney-07	-	-	-	28.61	1.8×10^3	4.7	8.5×10^3	1:4
	284-tonsil-07	+	+	-	28.54	4.9×10^3	1.8	8.9×10^3	-
	284-lymph node-07	-	+	+	30.08	4.8×10^2	6.5	3.1×10^3	-
Koslov	285-spleen-09	-	-	+	26.13	7.1×10^3	9.8	7.0×10^4	1:4
	285-kidney-09	-	-	-	32.30	1.9×10^2	3.6	6.9×10^2	-
	285-tonsil-09	-	-	+	27.86	8.8×10^3	2.4	2.1×10^4	1:4
	285-lymph node-09	-	-	+	29.79	4.6×10^2	12.5	5.7×10^3	1:4

[a] = Catalogue numbers deposited in the virus database of the EU Reference Laboratory for Classical Swine Fever Virus;

[b] = Dilution of transfected cells with a positive result for virus recovery;

+ = virus isolation from sample material was positive;

- = virus isolation from sample material was not possible

2.4 Virus isolation

Virus isolation was performed according to the standard protocols [4, 5]. Briefly, organ samples were homogenized and two concentrations of the organ suspension (undiluted and 1:10 diluted) were incubated on the cells. After two passages cell culture supernatant was collected and the cells were fixed by heat treatment (80°C for 3 h). For detection of viral antigen an indirect immune-peroxidase staining was performed using an NS3-specific monoclonal mouse antibody BVD/C16 ([27], dilution 1:25 in PBS-0.01%Tween) and a polyclonal rabbit anti-mouse horseradish peroxidase conjugate (dilution 1:200 in PBS-0.01% Tween, Catalogue no.: P0260, DAKO, Denmark).

2.5 Transfection of RNA

Electroporation of SK6 cells was performed as previously described with slight modifications [17]. Briefly, confluent SK6 cells, which were grown in a 10 cm dish, were resuspended in 0.4 ml PBS without Ca^{2+} and Mg^{2+}. For transfection, SK6 cells were mixed with RNA and added to a 0.2 cm cuvette before the pulse [950 µF and 180 V; Gene Pulser II (Bio-Rad, Germany)]. For the first set of samples different amounts of total RNA were used for transfection, ranging from approximately 1 µg to 33 µg of total RNA (corresponding to 10–15 µl; Table 1). Transfected cells were counted and diluted in 10-fold serial dilutions starting with a 1:4 dilution. They were seeded together with 1.5–2.0 x 10^6 naïve cells in six-well dishes. For RNA prepared from field samples a constant amount of 1 µg was used for electroporation (Table 2). Three days post transfection cells were fixed by heat treatment (80°C for 3 h). CSFV replication was detected by an indirect immunofluorescence assay using the NS3-specific monoclonal mouse antibody BVD/C16 (dilution 1:25 in PBS-0.01%Tween; incubation at 37°C for 2 h), and a goat anti-mouse IgG-Cy3 (Catalogue no.: 115-165-003, Dianova, Germany, dilution 1:800 in PBS-0.02% Tween; incubation at 37°C for 1 h). After each incubation step the cells were washed three times with PBS-0.05% Tween. Finally the wells were filled with ultrapure water and immunofluorescence was analyzed by an immunofluorescence microscope (Leica DFC420C, Germany).

Table 2. Conventional virus isolation and virus recovery after RNA electroporation testing field samples.

Sample ID	Virus isolation			qRT-PCR		RNA transfection of SK6 cells	
	PK15 cells	PK15(A) cells	STE cells	C_q value	copies /µg total RNA	Genome equivalents [copies/ reaction]	Recovery of infectious virus
154–1	+	+	+	24.35	3.2×10^4	3.2×10^4	+
154–2	+	-	+	22.56	9.3×10^4	9.3×10^4	+
154–3	+	-	+	21.45	5.0×10^5	5.0×10^5	+
154–4	+	-	-	21.35	1.5×10^6	1.5×10^6	+
154–5	+	+	-	27.90	1.5×10^4	1.5×10^4	+
154–6	-	-	-	28.43	1.5×10^3	1.5×10^3	+
154–7	-	-	+	27.38	1.2×10^4	1.2×10^4	+
154–8	+	+	+	20.83	4.9×10^5	4.9×10^5	+

+ = virus isolation from sample material was positive;

- = virus isolation from sample material was not possible

2.6 Ethics statement

The animal study followed the animal welfare regulations and standards according to EU Directive 201/63/EU and institutional guidelines. It was approved by the Specialized Department of Animal Welfare Service of the LAVES (Permit Number: LAVES AZ 11/0565) including recommendations of the ethics committee which was appointed by the LAVES according to the German animal welfare act (Tierschutzgesetz, § 15, Abs. 1, Nr. 2). Herewith we confirm that this ethic committee is the competent ethics authority and that no other specific Institutional Animal Care and Use Committee for the performance of the animal experiment were required. On the basis of a CSF clinical signs scoring system [28] the general health of study animals was recorded at least twice daily from three days before inoculation until completion of the study by experienced animal husbandry personnel and veterinarians. If the animals were in moribund stage they were euthanized by an intravenous overdose of pentobarbital followed by exsanguination (heart puncture) to minimize suffering. Since the clinical picture of animals infected with the CSFV strain Koslov is described in several publications, the number of animals of the control group was reduced. In addition, sample material (e.g. serum and organ samples) of the animals was used for teaching purposes and stored in the EURL reference material archive for further use in international ring trials.

2.7 Experimental design of the animal study

The animal experiment was carried out at the high containment unit of the EURL (Hannover, Germany). Eight pigs (37–47 days old; 12–15 kg) of both genders were purchased from a breeding farm of the University of Veterinary Medicine Hannover, Germany. They were tested negative for pestivirus genomes and for neutralizing antibodies against pestiviruses and were randomly allocated into two groups comprising five (RNA inoculation group, animal no: 355 to 359) and three animals (infection control group, animal no: 388 to 390), respectively. The animals were housed in two separate fenced pens under high containment conditions with no direct and indirect contact. They were kept litter-less, were fed with commercial pig feed and had access to water *ad libitum*. Animals of the RNA inoculation group were oronasally inoculated with 2 ml EMEM containing 3.3 µg RNA which was prepared from whole blood of an animal infected with highly virulent CSFV strain Koslov. RNA transfection of SK6 cells and subsequent serial dilution of the transfected cells was used to calculate the infection equivalents and to confirm that the RNA was infectious after transfection of susceptible cells. The infection equivalents were determined on the basis of the highest dilution of transfected cells with a positive result for virus recovery and the RNA amount used for inoculation of one animal. Accordingly, each animal was inoculated with ~1.3 x 10^3 infection equivalents. Additionally, to show that material used for RNA preparation is *per se* infectious, three weaner pigs were oronasally inoculated with the original material (control group). The inoculation volume was equal to the volume used for RNA preparation and corresponds to the infection dose of 1.13 x 10^5 $TCID_{50}$/ animal.

Rectal temperatures and clinical signs (according to Mittelholzer et al. 2000 with slight modifications, [28]) of all weaner pigs were recorded from three days before inoculation until euthanasia of the animals. Slight modifications were made for scoring parameters 9 "Defecation" and 10 "Leftovers in feeding trough". The criterion mild diarrhea was added to parameter 9 "Defecation" and scored with one point. Parameter 10 "Leftovers in feeding trough" could not be recorded for each animal because the animal groups were housed in one pen with one trough. Thus, the possible total number of scoring points per animal was lower than for the scoring scheme according to Mittelholzer et al., 2000. On day 1, 2, 3, 5, 7, 10, 14, 17 and 21 post inoculation blood samples (full blood and anti-coagulated EDTA blood) as well as nasal and pharyngeal swabs were taken and subjected to virological (qRT-PCR) and serological testing (virus neutralization assay). Leucocyte and thrombocyte counts were determined using a hematology analyzer (Abacus Junior vet/130464, Guder Medizintechnik, Germany). On the day of euthanasia, each animal was subjected to post-mortem examination. Organ samples (tonsil, lymph nodes, spleen, and kidney) were collected and subjected to virological testing (qRT-PCR). The ARRIVE (Animal Research: Reporting of *In Vivo* Experiments) Guidelines Checklist of this animal experiment is available as supporting information (S1 File).

2.8 Virus neutralization assay

Virus neutralization assay was performed with serum samples that were taken at seven days post inoculation and later on. The assay was performed according to the standard protocols [4, 5].

3. RESULTS

3.1 Virus rescue from experimentally derived organ samples

Sixteen organ samples from four experimentally infected animals were analyzed by a CSFV-specific qRT-PCR to quantify the viral genome load of each individual sample. All samples were tested positive for CSFV genome. The CSFV genome loads ranged between 1.9×10^2 and 5.8×10^5 genome equivalents/ µg RNA [quantification cycle (C_q) values from 20 to 30] (Table 1). Conventional virus isolation using PK15 cells was successful from ten out of 16 samples. The use of two additional cell lines resulted in virus isolation from 11 out of 16 [PK15(A) cells] and from 13 out of 16 (STE cells) samples, respectively. With regard to the four samples taken from Koslov infected pigs, virus isolation was more efficient on STE cells than on PK15 and PK15(A) cells. In routine CSFV diagnosis differences in efficiency of virus isolation on various CSFV susceptible cell lines are known. To our knowledge, those differences have not been systematically analyzed. Therefore, more than one cell line is usually used for virus isolation. For two samples (284-kidney and 285-kidney) virus isolation was not possible. Genome loads of these samples were 1.8×10^3 (284-kidney) and 1.9×10^2 (285-kidney) genome equivalents/ µg RNA, respectively (Table 1). To compare virus isolation with virus recovery after RNA transfection, total RNA from the same organ material previously subjected to qRT-PCR analysis (see above) was transfected into SK6 cells by electroporation. Emergence of virus was observed for 11 out of 16 samples. Remarkably, for sample 284-kidney recovery of infectious virus was possible only after RNA transfection whereas this sample was tested negative in conventional virus isolation on three different susceptible cell lines (Table 1). In general, the transfection of more than 1.9×10^5 genome equivalents was highly efficient. In this case, most of the samples were still positive after diluting the transfected cells 1:400 or 1:4,000 in suspensions of naïve cells. For six samples even transfection of total cellular RNA containing 5.7×10^3 to 7.0×10^4genome equivalents allowed virus recovery. In summary, virus rescue by RNA transfection was successful from each of the four experimentally infected pigs (Table 1).

Despite the use of antibiotics, microbial contamination was a problem for inoculation of homogenized organ samples on cells. For three out of 16 organ samples bacterial contamination of cells was observed during first passage of virus isolation. After transfection of SK6 cells with RNA isolated from the same samples CSFV was recovered without facing any problems with microbial contamination.

3.2 Virus rescue from field samples

To test whether RNA transfection can also be used as a general diagnostic tool, eight organ suspension samples taken during a CSFV outbreak in Serbia in 2010 were tested. Applying conventional virus isolation, six out of eight samples were tested positive on PK15 cells. Inoculation of these samples on two additional cell lines resulted in virus recovery from three [PK15(A) cells] and five (STE cells) samples, respectively. The CSFV genome copy numbers of these samples ranged between 1.5 x 10^3 and 1.5 x 10^6 copies/ µg RNA ($C_q$values from 21 to 28) (Table 2). Since it was shown that electroporation using ~1 µg total RNA was efficient, SK6 cells were transfected with 1 µg total RNA. Virus recovery was successful for all samples using one single cell line for transfection (Table 2). In comparison to this, three different cell lines were needed to rescue infectious virus by inoculation of organ material on susceptible cells for most of the samples. As already described for the experimentally derived organ samples, microbial contamination was also observed after inoculation of cells with suspension of field samples. For three out of eight organ samples at least one virus isolation set-up (concentrated organ homogenate or 1:10 dilution of organ homogenate) was contaminated during first virus passage. After transfection of SK6 cells with isolated RNA none of the cell cultures were affected by bacterial contamination. Furthermore, E2 coding sequences of two rescued viruses (sample no. 154-1 and 154-8) were 100% identical to the respective sequences obtained from the RNA of the original sample material.

3.3 Oronasal inoculation of pigs with RNA from infectious sample material

To answer the question whether RNA isolated from infectious sample material, can be regarded as non-dangerous goods in case of shipment an animal experiment was performed. All five animals which were inoculated with RNA prepared from blood of an animal infected with the highly virulent strain Koslov showed no clinical signs referred to CSFV infection. A slight increase of leucocyte counts was detected after RNA inoculation which reached a plateau (~25 x 10E9/ l) on five days post inoculation (Fig 1A). No leucopenia was observed and thrombocyte counts (325 x 10E9/ l - 715 x 10E9/ l) were in physiological range except for one animal (animal no.: 356) during the first three days after inoculation (Fig 1C). Analysis of serum samples taken on day 7, 10, 14, 17 and 21 post inoculation by virus neutralization assay demonstrated the absence of CSFV-specific neutralizing antibodies. Furthermore, swabs, leucocytes and organ samples (lymph node, tonsil, spleen and kidney) were tested negative for CSFV genome. One animal (animal no.: 357) showed a slightly increased body temperature (40°C) between five and eight days post inoculation. This animal showed mild diarrhea for two days on four (clinical score = 1) and five (clinical score = 0.5) days post inoculation (Fig 2A). However, samples taken from this animal were all tested negative for CSFV genome or antibodies.

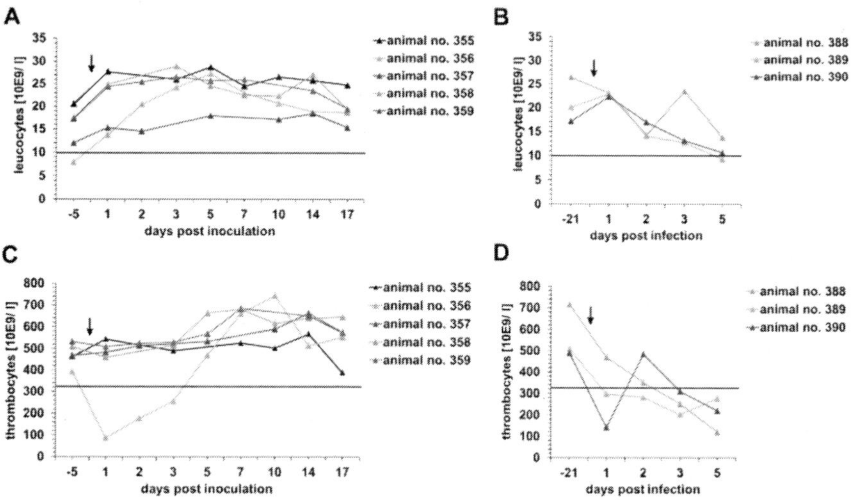

Figure1. Leucocyte and thrombocyte counts of animals inoculated with RNA or infected with CSFV Koslov. Anti-coagulated EDTA blood samples were taken 21 (control group) or 5 days (RNA inoculation group) before inoculation of the animals and on day 2, 3, 5, 7, 10, 14 and 17 after inoculation. Inoculation or infection of the animals is marked by an arrow. Leucocyte (A and B) and thrombocyte counts (C and D) were determined using hematology analyzer. Black lines within the graphs mark the cut-off value for leucopenia or thrombocytopenia, respectively. Due to ethical reasons the animals of the control group were also part of a second study and therefore inoculated with CSFV Koslov 21 days after initial sampling.

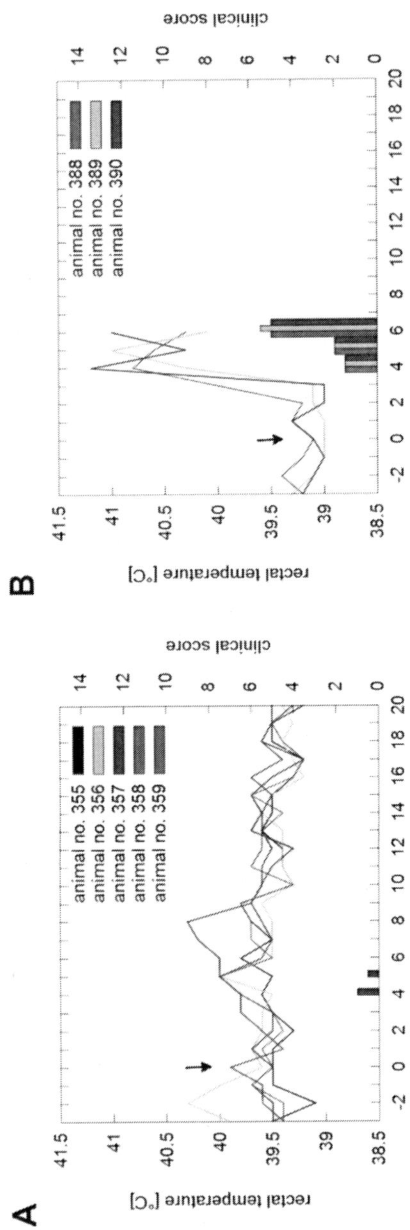

Figure 2. Rectal temperatures and clinical scores of animals inoculated with RNA or infected with CSFV Koslov. Temperatures and clinical signs of the RNA inoculation (A; animal no.: 355 to 359) and infection control group (B; animal no.: 388 to 390), respectively, were recorded daily starting three days prior inoculation. Inoculation or infection of the animals is marked by an arrow. Body temperatures ≥ 40.0°C for two consecutive days were considered to be fever.

The RNA used for animal inoculation was only infectious after transfection of SK6 cells. Direct inoculation of SK6 cells with RNA mixed in culture medium lead to no virus production whereas virus was detected for an included positive control (inoculation of the cells with infectious cell culture supernatant containing CSFV Alfort/187).

In contrast to the RNA inoculation group, animals of the infection group developed high fever on day three post infection, showed severe clinical signs of CSF, and were finally euthanized on day seven post infection in a moribund stage (Fig 2B). A decrease of leucocyte and thrombocyte counts was observed after infection (Fig 1B and 1D). Since the animals were euthanized at an early time point after infection, the presence of virus neutralizing antibodies was not expected and therefore not determined. In contrast to the RNA inoculation group, low CSFV genome loads were already detectable in leucocytes taken on day two post inoculation (C_q values 34.98 to 37.41), followed by an increase until last sampling day (C_q values 21.08 to 25.82). Swabs were tested positive on day three post infection (C_q values 38.17 to 38.3) until last sampling day (C_q values 29.17 to 32.76). Furthermore, high copy numbers of CSFV RNA were detected analyzing the organ samples of the control group (C_q values 20.9 to 31.29).

4. DISCUSSION

Virological methods for the diagnosis of CSF are based on virus isolation after incubation of sample material on susceptible cells and detection of virus antigen and virus genome [4]. RNA transfection methods may represent a good alternative to rescue positive stranded RNA viruses from samples that are negative in virus isolation [20–23]. So far, knowledge on efficiency of virus rescue by RNA transfection directly compared to conventional virus isolation using samples of CSFV infected pigs is still missing. To conclude whether RNA transfection can be recommended in addition to conventional virus isolation, the present study focused on direct comparison of both methods using organ samples of experimentally and field infected pigs, respectively. It was demonstrated that conventional virus isolation was time and labor intensive as for several samples more than one cell line had to be used for successful virus isolation. RNA transfection was successful in particular for critical samples that were only positive for one out of three tested cell lines. For two samples (284-kidney, 154–6), rescue of infectious virus was possible only after RNA transfection but not with the conventional virus isolation, even when three different cell lines were inoculated. However, virus rescue by RNA transfection showed also limitations which might be due to the lack of integrity of the 12.3 kb CSFV RNA genome after RNA preparation. In previous studies similar observations have been reported [20, 21, 23]. Since determination of CSFV genome copy numbers is based on amplification of a short genome region, this method cannot provide information on the actual number of infectious viral genomes. Nevertheless, it can be assumed that the amount of intact CSFV genomes increases when high amounts of genome equivalents are used for transfection. Furthermore, efficiency of RNA transfection for recovery of virus is dependent on several other factors, including amount of RNA, sample quality, and cell viability after transfection. The results of our study showed, that virus rescue after transfection of 1.9×10^5 CSFV genome equivalents was very efficient. For these samples, infectious virus was recovered even in high dilutions of transfected cells. Genome equivalents in this range can be expected in 10 to 15 µl volumes of RNA preparations with C_q values between 20 and 24 in qRT-PCR. However, it was also shown that samples with higher C_q values (25 to 30) allow virus rescue by RNA transfection. The minimal amount of extracted RNA resulting in successful virus rescue was tested using field samples from an outbreak in Serbia. One microgram of total RNA, which contained between 1.5×10^3 and 1.5×10^6 CSFV genome equivalents, was sufficient for successful virus rescue of all field samples. From the methodological point of view, RNA transfection is less labor and time intensive and not prone to microbial contamination of cell culture. It is well known that microbial contamination can significantly limit the efficiency of virus isolation after inoculation of cells with infectious sample suspension. In this context, it seems to be likely that bacterial contamination observed for sample 284-kidney was at least partly responsible for the failure of conventional virus isolation although virus rescue by RNA transfection was successful. Furthermore, it is known that autolyzed sample material has cytotoxic effects on cell culture as well [4]. Insufficient quality is often a problem of diagnostic samples obtained from wild boar and in such cases RNA transfection might be a valuable alternative. In a recently published study on FMDV it was also described that RNA transfection represents a good alternative for virus rescue when conventional virus isolation is not possible [21]. Thus it can be concluded, that RNA transfection may represent a valuable tool for generation and isolation of other positive stranded RNA viruses as well.

For a clear classification of RNA as non-hazardous material the innocuity of RNA isolated from samples of CSFV infected pigs, has to be assessed in the natural host. To our knowledge inoculation studies of pigs or wild boar via oronasal route with prepared RNA have not been reported before. The results of the present study show that experimental oronasal inoculation of RNA from the highly virulent CSFV strain Koslov did not result in infection or disease of pigs. Since CSFV Koslov causes a severe acute disease with 100% lethality, this strain has been frequently used as challenge virus and therefore is an appropriate read-out system to proof the non-hazardous character of the RNA preparations used for inoculation of pigs [29–33]. As expected, severe clinical signs were observed for the animals of the infected control group and all animals had to be euthanized in a moribund stage on day seven post infection. Absence of clinical signs, viral genomes and CSFV specific antibody response gave no hints that CSFV RNA might be *per se* infectious and hazardous for pigs after oronasal exposure. Since neither CSFV genomes nor CSFV-specific antibodies were present in any of the samples, it can be concluded that the mild unspecific clinical signs observed for animal 357 and the transient decrease of thrombocyte counts after RNA inoculation detected for animal 356 were not due to CSFV infection. In the present study pigs were inoculated with 3.3 µg of RNA corresponding to a dose of $\sim 10^3 TCID_{50}$ produced after electroporation of susceptible cells. Similar amounts of RNA material are likely to be handled and transported between laboratories. To our knowledge, the minimal infectious dose of highly virulent CSFV strain Koslov has not been determined so far. However, previous studies have shown that an inoculation dose between $\sim 10^{2.5}$ and $10^3 TCID_{50}$/ animal of virulent CSFV strains (Brescia and Alfort/187) resulted in animal infection and disease [34, 35] and a dose of 200 $TCID_{50}$/ animal was used for challenge infection with CSFV Koslov [36]. Based on these facts, a dose of 10^3 $TCID_{50}$/ animal of the highly virulent strain Koslov can be expected to be above the minimal infectious dose. Taken into account the amount and quality of RNA used for the inoculation of pigs described in the present study together with the assumption that exposure of pigs with RNA samples represents a very rare event, it seems to be very unlikely that an accidental oral up-take of RNA sample will lead to infection of pigs.

Biosecurity classification during transportation of samples containing full length genomes of other positive stranded RNA has to be estimated for each RNA virus separately and depends also on legislation of the country. For example in the United Kingdom samples containing full length genome of FMDV require the same biosecurity level as described for infectious sample material. This includes that only shipment to laboratories with appropriate containment facilities is allowed [37].

Taken together, RNA transfection is a highly efficient approach for obtaining infectious CSFV isolates. It represents a valuable alternative to conventional virus isolation in particular for samples from which infectious virus cannot be isolated or when infectious material is not available. Accordingly, it is recommended to include the method of virus rescue by RNA transfection as an alternative approach for virus isolation in the Diagnostic Manual for CSF (Commission Decision 2002/106/EC). Furthermore, this study revealed that RNA prepared from samples of CSFV infected pigs, at least for the RNA amount used in the animal experiment described here, is *per se* not infectious for the host species and can therefore be classified as non-hazardous material. Nevertheless, the presented data highlighted that RNA can be converted relatively easy into hazardous material after transfection. Appropriate security arrangements to prevent intentional misuse or unintentional contamination (e.g. transfection reagents) are still required, although these security arrangements may not be the same as for infectious virus. The RNA transfection method described here can contribute to facilitate the exchange of virus isolates among national and international reference laboratories for CSFV as well as other authorized laboratories.

ACKNOWLEDGMENTS

We thank Alexandra Meindl-Böhmer for her contributions in the initial stage of the project and during the animal experiment. For animal care, we thank Günter Thiem, Monika Berg and Holger Mosch. We are grateful to Alexandra Beckmann, Inga Grotha, Benjamin Ostermann, Gabriele Müller and Karin Ruthenberg for their excellent technical assistance.

AUTHOR CONTRIBUTIONS

Conceived and designed the experiments: DM SS AP PB. Performed the experiments: DM SS AP. Analyzed the data: DM SS AP PB. Contributed reagents/materials/analysis tools: DM SS AP PB. Wrote the paper: DM AP PB.

REFERENCES

1. Simmonds P, Becher P, Collett MS, Gould EA, Heinz FX, Meyers G, et al. (2011) Family Flaviviridae. In: King A, Lefkowitz E, Adams M J and Carstens E B, editors. Virus taxonomy: Ninth Report of the International Committee on Taxonomy of Viruses. Academic Press, San Diego, CA. pp. 1003–1020.

2. Vandeputte J, Chappuis G (1999) Classical swine fever: the European experience and a guide for infected areas. Revue scientifique et technique (International Office of Epizootics) 18: 638–647. pmid:10588007

3. Edwards S, Fukusho A, Lefevre PC, Lipowski A, Pejsak Z, Roehe P, et al. (2000) Classical swine fever: the global situation. Veterinary microbiology 73: 103–119. pmid:10785321

4. Anonymous (2002) Commission Decision of February 2002 approving a Diagnostic Manual establishing diagnostic procedures, sampling methods and criteria for evaluation of the laboratory tests for the confirmation of classical swine fever (2002/106/EC). Official Journal of the European Communities: 39–71.

5. Anonymous (2007) EU Diagnostic Manual for Classical swine fever (CSF) Diagnosis: Technical Part (Third Draft June 2007). Available: http://www.tiho-hannover.de/kliniken-institute/institute/institut-fuer-virologie-zentrum-fuer-infektionsmedizin/eu-and-oie-reference-laboratory/downloads/. EU and OIE Reference laboratory for Classical Swine Fever Virus. Accessed 12 November 2014.

6. Greiser-Wilke I, Depner K, Fritzemeier J, Haas L, Moennig V (1998) Application of a computer program for genetic typing of classical swine fever virus isolates from Germany. J Virol Methods 75: 141–150. pmid:9870589

7. Greiser-Wilke I, Zimmermann B, Fritzemeier J, Floegel G, Moennig V (2000) Structure and presentation of a World Wide Web database of CSF virus isolates held at the EU reference laboratory. Veterinary microbiology 73: 131–136. pmid:10785323

8. Dreier S, Zimmermann B, Moennig V, Greiser-Wilke I (2007) A sequence database allowing automated genotyping of Classical swine fever virus isolates. J Virol Methods 140: 95–99. pmid:17187869

9. Postel A, Schmeiser S, Bernau J, Meindl-Boehmer A, Pridotkas G, Dirbakova Z, et al. (2012) Improved strategy for phylogenetic analysis of classical swine fever virus based on full-length E2 encoding sequences. Vet Res 43: 50.

10. Postel A, Moennig V, Becher P (2013) Classical swine fever in Europe—the current situation. Berliner und Munchener tierarztliche Wochenschrift 126: 468–475. pmid:24511821

11. Poole TL, Wang C, Popp RA, Potgieter LN, Siddiqui A, Collett MS (1995) Pestivirus translation initiation occurs by internal ribosome entry. Virology 206: 750–754. pmid:7831836

12. Rijnbrand R, van der Straaten T, van Rijn PA, Spaan WJ, Bredenbeek PJ (1997) Internal entry of ribosomes is directed by the 5' noncoding region of classical swine fever virus and is dependent on the presence of an RNA pseudoknot upstream of the initiation codon. J Virol 71: 451–457. pmid:8985370

13. Meyers G, Tautz N, Becher P, Thiel HJ, Kummerer BM (1996) Recovery of cytopathogenic and noncytopathogenic bovine viral diarrhea viruses from cDNA constructs. J Virol 70: 8606–8613. pmid:8970985

14. Meyers G, Thiel HJ, Rumenapf T (1996) Classical swine fever virus: recovery of infectious viruses from cDNA constructs and generation of recombinant cytopathogenic defective interfering particles. J Virol 70: 1588–1595. pmid:8627678

15. Ruggli N, Tratschin JD, Mittelholzer C, Hofmann MA (1996) Nucleotide sequence of classical swine fever virus strain Alfort/187 and transcription of infectious RNA from stably cloned full-length cDNA. J Virol 70: 3478–3487. pmid:8648680

16. Pankraz A, Thiel HJ, Becher P (2005) Essential and nonessential elements in the 3' nontranslated region of Bovine viral diarrhea virus. J Virol 79: 9119–9127. pmid:15994806

17. Gallei A, Blome S, Gilgenbach S, Tautz N, Moennig V, Becher P (2008) Cytopathogenicity of Classical Swine Fever virus correlates with attenuation in the natural host. J Virol 82: 9717–9729.

18. Lamp B, Riedel C, Roman-Sosa G, Heimann M, Jacobi S, Becher P, et al. (2011) Biosynthesis of classical swine fever virus nonstructural proteins. J Virol 85: 3607–3620.

19. Austermann-Busch S, Becher P (2012) RNA structural elements determine frequency and sites of nonhomologous recombination in an animal plus-strand RNA virus. J Virol 86: 7393–7402.

20. Belsham GJ, Jamal SM, Tjornehoj K, Botner A (2011) Rescue of foot-and-mouth disease viruses that are pathogenic for cattle from preserved viral RNA samples. PLoS One 6: e14621.

21. Bisht P, Mohapatra JK, Subramaniam S, Das B, Pande V, Biswal JK, et al. (2014) Efficient rescue of foot-and-mouth disease virus in cultured cells transfected with RNA extracted from clinical samples. J Virol Methods 196: 65–70.

22. Fowler VL, Bankowski BM, Armson B, Di Nardo A, Valdazo-Gonzalez B, Reid SM, et al. (2014) Recovery of Viral RNA and Infectious Foot-and-Mouth Disease Virus from Positive Lateral-Flow Devices. PLoS One 9: e109322.

23. **23.**Hofmann MA, Thur B, Liu L, Gerber M, Stettler P, Moser C, et al. (2000) Rescue of infectious classical swine fever and foot-and-mouth disease virus by RNA transfection and virus detection by RT-PCR after extended storage of samples in Trizol. J Virol Methods 87: 29–39. pmid:10856750

24. Grummer B, Fischer S, Depner K, Riebe R, Blome S, Greiser-Wilke I (2006) Replication of classical swine fever virus strains and isolates in different porcine cell lines. DTW Deutsche tierarztliche Wochenschrift 113: 138–142. pmid:16716048

25. Kasza L, Shadduck JA, Christofinis GJ (1972) Establishment, viral susceptibility and biological characteristics of a swine kidney cell line SK-6. Research in veterinary science 13: 46–51. pmid:4336054

26. Hoffmann B, Beer M, Schelp C, Schirrmeier H, Depner K (2005) Validation of a real-time RT-PCR assay for sensitive and specific detection of classical swine fever. J Virol Methods 130: 36–44. pmid:16055202

27. Greiser-Wilke I, Dittmar KE, Liess B, Moennig V (1992) Heterogeneous expression of the non-structural protein p80/p125 in

cells infected with different pestiviruses. J Gen Virol 73 (Pt 1): 47–52. pmid:1309861

28. Mittelholzer C, Moser C, Tratschin JD, Hofmann MA (2000) Analysis of classical swine fever virus replication kinetics allows differentiation of highly virulent from avirulent strains. Veterinary microbiology 74: 293–308. pmid:10831853

29. Kaden V, Lange B (2001) Oral immunisation against classical swine fever (CSF): onset and duration of immunity. Veterinary microbiology 82: 301–310. pmid:11506924

30. Kaden V, Schurig U, Steyer H (2001) Oral immunization of pigs against classical swine fever. Course of the disease and virus transmission after simultaneous vaccination and infection. Acta virologica 45: 23–29. pmid:11394574

31. Reimann I, Depner K, Utke K, Leifer I, Lange E, Beer M (2010) Characterization of a new chimeric marker vaccine candidate with a mutated antigenic E2-epitope. Veterinary microbiology 142: 45–50.

32. Blome S, Aebischer A, Lange E, Hofmann M, Leifer I, Loeffen W, et al. (2012) Comparative evaluation of live marker vaccine candidates "CP7_E2alf" and "flc11" along with C-strain "Riems" after oral vaccination. Veterinary microbiology 158: 42–59.

33. Blome S, Gabriel C, Schmeiser S, Meyer D, Meindl-Bohmer A, Koenen F, et al. (2014) Efficacy of marker vaccine candidate CP7_E2alf against challenge with classical swine fever virus isolates of different genotypes. Veterinary microbiology 169: 8–17.

34. Dahle J, Liess B (1995) Comparative study with cloned classical swine fever virus strain Alfort and Glentorf: clinical, pathological, virological and serological findings in weaner pigs. Wien Tierarztl Monatsschr 82: 232–238.

35. Weesendorp E, Backer J, Stegeman A, Loeffen W (2009) Effect of strain and inoculation dose of classical swine fever virus on within-pen transmission. Vet Res 40: 59.

36. Kaden V, Lange E, Steyer H, Lange B, Klopfleisch R, Teifke JP, et al. (2008) Classical swine fever virus strain "C" protects the offspring by oral immunisation of pregnant sows. Veterinary microbiology 130: 20–27.

37. Anonymous (2001) Anti-terrorism, Crime and Security Act 2001, Chapter 24. Available: http://www.legislation.gov.uk/ukpga/ 2001/ 24/contents. The Stationery Office Limited under the authority and superintendence of Carol Tullo, Controller of Her Majesty's Stationery Office and Queen's Printer of Acts of Parliament. Accessed: 4 November 2014.

CHAPTER 14

Contrasting clinical outcomes in two cohorts of cats naturally infected with feline immunodeficiency virus (FIV)

Paweł M. Be̜czkowski a,b, *, Annette Litster c , Tsang Long Lin d , Dominic J. Mellor e , Brian J. Willett a , Margaret J. Hosie a

a MRC Centre for Virus Research, University of Glasgow, Glasgow, UK

b Small Animal Hospital, University of Glasgow, Glasgow, UK

cDepartment of Veterinary Clinical Sciences, Purdue University, West Lafayette, IN 47907, USA

d Indiana Animal Disease Diagnostic Laboratory and Department of Comparative Pathobiology, Purdue University, West Lafayette, IN 47907, USA

e School of Veterinary Medicine, University of Glasgow, Glasgow, UK

ABSTRACT

Despite over 25 years of feline immunodeficiency virus (FIV) research, relatively little is known about the longitudinal course of FIV infection following natural infection. In contrast to published reports of experimental infections using lethal strains of the virus, clinical signs of naturally acquired FIV infection can be mild or inapparent, rather than life-threatening. In this prospective, longitudinal controlled study, based in Chicago, IL ($n = 17$) and Memphis, TN ($n = 27$), we investigated two cohorts of privately owned, naturally infected cats kept under different housing conditions. Cats in the Chicago cohort (Group 1) were kept in households of ≤2 cats, while the Memphis cohort (Group 2) comprised part of a large multi-cat household of over 60 cats kept indoors only, with unrestricted access to one another.

The majority of cats from Group 1 did not display clinical signs consistent with immunodeficiency during the 22-month observation period. In contrast, the outcome of infection in Group 2 was dramatically different; 17/27 (63%) of cats lost a median of 51.3% of their bodyweight ($P < 0.0005$) and died during the study period, with lymphoma being the most common cause of mortality.

Although the decrease in CD4+ T cell count between enrolment and terminal disease was significant ($P = 0.0017$), the CD4:CD8 ratio at the time of enrolment did not reliably distinguish FIV-positive cats classified as 'healthy' and 'not healthy' at either cohort. FIV load at enrolment was significantly lower in Group 1 than in Group 2 ($P < 0.0001$), but there were no significant differences at enrolment between healthy and not healthy cats at either group.

In conclusion, the results of this study suggest that management and housing conditions impact on disease progression and survival times of FIV-positive cats.

KEYWORDS

FIV; Natural infection; Clinical outcome; FIV load; CD4:CD8; Lymphoma

1. INTRODUCTION

Feline immunodeficiency virus (FIV) is an important pathogen of domestic cats that is distributed worldwide. Prevalence estimates vary from 1% to 44%, depending on the geographical location as well as the entry criteria of each study (Hartmann, 1998 and Ishida et al., 1989). Outdoor, adult, male, sick cats with bite wounds are at increased risk of infection, since transmission is mainly via bites from infected cats (Hartmann, 1998). Clinical signs are similar to those described for human immunodeficiency virus-acquired immune deficiency syndrome (HIV-AIDS) and include signs of immunosuppression, hematopoietic changes and neoplasia (Hartmann, 2011). The progression of experimental FIV infection can be monitored using the staging system used in HIV infection: acute, persistent generalized lymphadenopathy, asymptomatic carrier phase (AC), AIDS-related complex (ARC) and AIDS (English et al., 1994). While this classification is useful in well-controlled, experimental settings, it has limitations under field conditions, mainly because of strain-related variability in pathogenicity, variation in the ages of cats at the time of infection as well as possible exposure to secondary pathogens (Hartmann, 2012, Pedersen et al., 2001 and Podell et al., 1997). Hence, there have been variable morbidity and mortality rates reported in naturally infected cats (Addie et al., 2000, Hofmann-Lehmann et al., 1997, Kohmoto et al., 1998 and Liem et al., 2013).

Providing an accurate prognosis for FIV-positive cats is important so that optimal care can be provided by veterinarians and cat owners. Surrogate markers for HIV infection are well established and valuable in monitoring both responses to therapy and the likelihood of disease progression (Churchill, 1997, Katzenstein, 2003, Okonji et al., 2012, Shaunak and Teo, 2003 and Smith and Stein, 2002). In contrast, to our knowledge, prospective controlled studies investigating the course of FIV infection and markers of disease progression have not been described. Consequently, the longitudinal course of naturally acquired FIV infection is poorly described and the variability in rates of progression to immunodeficiency are poorly understood.

In this controlled study, we examined clinical and laboratory parameters from two cohorts of cats living in different locations and kept under different housing conditions. The aim of the study was to report clinical findings, post-mortem findings, bodyweight, CD4+ T cell counts, FIV load and phylogenetic classification of viral envelope genes in two groups of cats naturally infected with FIV.

2. MATERIALS AND METHODS

2.1. Cats

Forty-four neutered, FIV-positive (SNAP FIV/FeLV Combo Test (IDEXX Laboratories, Westbrook, MN) cats were enrolled, with an equal number of age- and sex-matched FIV-negative cats from the same geographical locations. In order to obtain an adequate sample size FIV-positive cats and matched control cats during the 2-year enrolment period, it was necessary to source cats from two separate locations. FIV-positive results were further confirmed by virus isolation (Hosie et al., 2009). All cats were feline leukaemia virus (FeLV) antigen negative at enrolment. The number of cats enrolled into the study was capped at 44 due to the capacity of the laboratory at which molecular analyses were performed.

Seventeen of the 44 FIV-positive cats enrolled (Group 1) had been previously adopted from a large metropolitan adoption-guarantee shelter (PAWS Chicago) and lived in single-cat households in Chicago, IL, USA except for seven cats: two cats (C7 and C4) cohabited in a two-cat household; one cat (C13) lived in a two-cat household with a FIV-negative cat enrolled in the study; one cat (C9) was housed at PAWS Chicago for the first 11 weeks of the study and then was adopted into a house with an FIV-positive cat not enrolled in the study; and three cats (C2, C15 and C21) were housed at PAWS Chicago in a room containing up to three FIV-positive cats before they were each adopted into single cat households at 2, 14 and 58 weeks after enrolment, respectively.

The remaining twenty-seven FIV-positive cats enrolled (Group 2) were housed together in a large multi-cat household operating as an FIV-positive cat rescue in Memphis, TN, where a total of 53 FIV-positive and 10 FIV-negative cats were housed indoors with unrestricted access to one another. None of 10 FIV-negative cats from the Memphis cat rescue died during the study period. Since these cats were not age- or sex-matched for any of the enrolled FIV-infected cats from the same household (Group 2), their health status was not monitored using the study protocol. All enrolled FIV-negative cats at both locations lived in households of three cats or fewer.

The study and its aims were reviewed and approved by University of Glasgow Ethics Committee and the Purdue Animal Care and Use Committee. Cat owners provided written informed consent for their participation in the study.

2.2. Study timeline and collection dates

At enrolment, all cats underwent a general physical examination by a registered specialist in feline medicine (AL) and blood was collected for determination of FIV status and laboratory analysis. Oral examinations were conducted as part of the physical examination and a gingival score (0–3/3; Lobprise, 2007) was assigned. At the time of enrolment, FIV-positive cats were classified as 'healthy' if there were no abnormalities on physical examination and their gingival score was 0/3 or 1/3. Cats were classified as 'not healthy' if at least one clinical abnormality was detected, or if their gingival score was 2/3 or 3/3. The clinical classification at enrolment was based solely on physical examination findings and remained with each cat for the 22-month study period, regardless of any further changes. An entry criterion for age- and sex-matched FIV-negative cats was healthy status at the enrolment examination.

Serial physical examinations and specimen collections were made at 6-monthly intervals in FIV-positive cats over the study period. Physical examinations and specimen collections were performed at enrolment and 12 months later in FIV-negative cats. The date of the first known positive FIV ELISA test was obtained for all FIV-positive cats and the time between first diagnosis and study enrolment (months) was calculated. Abnormalities detected during the clinical examinations are listed in Table 1.

2.3. Laboratory and post-mortem examinations

Flow cytometric analyses were performed as previously described (Lin and Litster, 2013). FIV loads were measured using a commercially available PCR test (IDEXX FIV RealPCR Test, IDEXX Laboratories, West Sacramento, CA). The assay detects the presence of viral nucleic acid, including both genomic DNA and viral RNA, in peripheral blood leukocytes with 80.5% sensitivity and 99.9% specificity (IDEXX Laboratories, West Sacramento, CA).

Post-mortem examinations were performed by a specialist veterinary pathologist (TLL).

Table 1. Clinical abnormalities detected on physical examination in FIV-positive cats ($n = 44$) over the 22-month study period.

Description	Number of cats with clinical abnormalities (%)		
	Group 1 ($n = 17$)	Group 2 ($n = 27$)	All FIV-positive cats ($n = 44$)
Dental/oral cavity (faucitis)	12 (71)	19 (70)	31 (70)
Skin (alopecia, dermatitis)	10 (59)	12 (44)	22 (50)
Ears	3 (18)	4 (15)	7 (16)
Ocular disease	2 (12)	5 (19)	7 (16)
Cardiovascular	3 (18)	2 (7)	5 (11)
Digestive tract	0 (0)	4 (15)	4 (9)
Respiratory	0 (0)	2 (7)	2 (5)
Lymph nodes	0 (0)	0 (0)	0 (0)
Urogenital	0 (0)	0 (0)	0 (0)
Neurologic	0 (0)	0 (0)	0 (0)
Musculoskeletal	0 (0)	0 (0)	0 (0)
Weight loss/anorexia	0 (0)	22 (81)	22 (50)

2.4. Phylogenetic analysis

Maximum likelihood (ML) phylogenetic trees were constructed in MEGA5 (Tamura et al., 2011) under the HKY nucleotide substitution model, selected through jMODELTEST (Posada, 2008). Statistical support for the ML trees was estimated using 1000 bootstrap replicates (Efron et al., 1996). Multiple *env* sequences ($n = 355$), from cats enrolled in the study were subjected to rigorous recombination testing as described previously (Bęczkowski et al., 2014).

2.5. Statistical analysis

Graphing and statistical data analyses were performed using commercially available software (GraphPad Prism version 5.00, GraphPad Software). Descriptive data were shown as medians and interquartile range (IQR; median, 5th and 95th quartiles). Given the relatively small sample size, and after inspection of the data distributions, Mann–Whitney and Wilcoxon matched pairs tests were used to test hypotheses regarding differences in laboratory parameters between and within cat groups. Binary data were analyzed using Fisher's exact test. FIV load data were tested for correlation using Spearman correlation tests. Kaplan–Meier curves were compared using the Mantel–Cox 'log-rank' test and tested with the log-rank test for trends. Significance was set at $P < 0.05$.

3. RESULTS

3.1. Cats

Table 2 provides breed, gender, age and time since first FIV-positive test data for all FIV-positive cats and for cats from Group 1 and Group 2. Each matching pair of FIV-positive and FIV-negative cats was aged within 2 years of each other. All enrolled cats remained FeLV-negative over the study period and none of the FIV-negative cats seroconverted.

Table 2. Breed, gender, age and age at first known positive feline immunodeficiency virus (FIV) ELISA test at enrolment.

	Group 1 ($n = 17$)	Group 2 ($n = 27$)	All FIV-positive cats ($n = 44$)
Breed	DSH (94%), DLH (6%)	DSH (74%), DLH (15%), Siamese X (11%)	DSH (82%), DLH (11%), Siamese X (7%)
Gender	FS (18%), MN (82%)	FS (33.3%), MN (66.6%)	FS (27%), MN (73%)
Age (years; median/range)	4/1–9	5/2–10	4.5/1–10
Age at first known positive FIV ELISA test (years; median/range)	2.3/0.2–7.8	1.6/0–4.6	1.8/0–7.8
Time from first known positive FIV ELISA test to enrolment (years; median/range)	1.3/0–3.3	3/0.1–8	2/0–8

DSH, domestic shorthair; DLH, domestic longhair; X, cross; FS, female spayed; MN, male neutered.

3.2. Health status

In the FIV-positive group at the time of enrolment, there were equal numbers of healthy (*n* = 22) and not healthy animals (*n* = 22). In Group 1, 10 cats were classified as healthy (59%) and seven cats were classified as not healthy (41%); all but one cat remained in this classification during the observation period. In Group 2, 12 cats were initially classified as healthy (44%) and 15 cats that were classified as not healthy (56%) remained in these classifications throughout the study. However, 63% of cats (17/27) from Group 2 experienced severe weight loss and died during the study period. Half (6/12) of the cats classified as healthy and 73% (11/15) of cats classified as not healthy died during the study ($P = 0.26$). Almost half of cats diagnosed at post-mortem with lymphoma (44%; 4/9) had been classified as healthy at the time of enrolment (Table 3).

3.3. Post-mortem findings

During the 22-month study period, 1/17 (5.9%) FIV-positive cats from Group 1 and 17/27 (63%) FIV-positive cats from Group 2 died. Fig. 1 illustrates a Kaplan–Meier survival plot of both cohorts. The cats that died were examined post-mortem and the findings are shown in Table 3. In Group 2, post-mortem data were not available for one cat (M33).

The most common pathological finding at post-mortem in Group 2 was lymphoma; various anatomical types were documented in 9/16 (56%) cats from Group 2 (Table 4). Four cats had lymphoma limited to a single site (retro-bulbar site in one case and bone marrow in three cases), while in five other cats, varying degrees of dissemination were observed. Bone marrow involvement was noted in all but one case. The second most common finding was nodal lymphoma, diagnosed in 5/9 cats (56%) and accompanied by splenic lymphoma in 3/9 cases (33%). One cat (M11) had a disseminated form involving the kidneys, liver, jejunum, heart, trachea and tongue. The median time since first FIV-positive ELISA test to death in cats with lymphoma was 5.3 years (range, 2.9–9 years).

Table 3. Health status on the day of enrolment and post-mortem diagnoses for FIV-positive cats deceased during the study (Group 1, n = 1; Group 2, n = 17).

Cat #	Health status at enrolment	Post-mortem diagnosis
M3	NH	Lymphoma – retrobulbar
M5	NH	Lymphoma – bone marrow
M10	H	Lymphoma – bone marrow and lymph nodes
M11	H	Lymphoma – multicentric
M14	NH	Lymphadenomegaly; bone marrow hyperplasia
M16	NH	Pulmonary congestion and oedema with focal bacterial bronchopneumonia
M25	H	Upper and lower bacterial respiratory infection, probably secondary to viral infection
M26	NH	Chronic focal granulomatous nematode larval pneumonia; bone marrow hyperplasia
M30	NH	Lymphoid depletion; Aelurostrongylus abstrusus pneumonia; left front leg suppurative dermocellulitis; bone marrow hyperplasia, with erythroid and myeloid hyperplasia
M31	NH	Lymphoma – bone marrow
M33	NH	No post-mortem performed
M41	NH	Lymphoma – bone marrow and lymph nodes
M44	NH	Emaciation/nephritis/pancreatitis
M46	H	Lymphoma – bone marrow
M48	NH	Lymphoma – bone marrow and lymph nodes
M49	H	Lymphoma – bone marrow
M50	H	Undetermined/emaciation
C2	H	Cardiac failure, hypertrophic cardiomyopathy

C, Chicago, Group 1; M, Memphis, Group 2; H, healthy on the day of enrolment (no abnormalities detected on a physical examination); NH, not healthy on the day of enrolment (at least one abnormality detected on a physical examination).

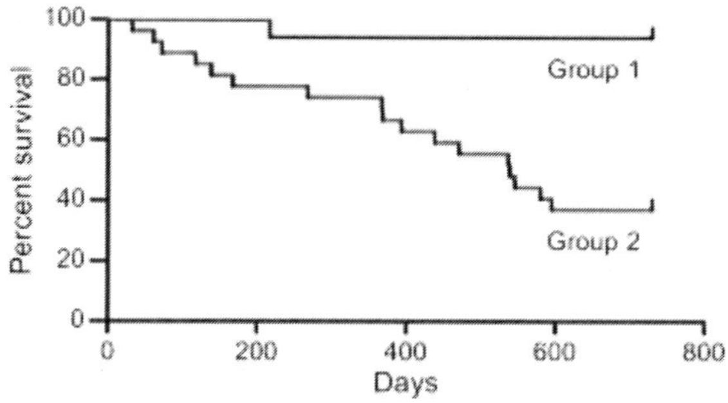

Figure 1. Kaplan–Meier survival plots for FIV-positive cats from Group 1 (black; n = 17) and Group 2 (red; n = 27). Survival rates in Group 1 and Group 2 over the 22-month study period were 94% and 37%, respectively (P = 0.0006, Gehan–Breslow–Wilcoxon test). (For interpretation of the references to colour in this figure legend, the reader is referred to the web version of this article.)

Lymphoid hyperplasia and lymphoid depletion were noted concurrently in 13/16 (81%) cats. These findings were documented in various tissues, mostly in the spleen and lymph nodes, but in four cats, Peyer's patch atrophy was also diagnosed. Erythroid bone marrow hyperplasia was documented in 9/16 cats (56%), as was myeloid bone marrow hyperplasia. Multifocal lymphocytic interstitial nephritis was diagnosed in six cats (37.5%). In four cats, there were pathological signs of respiratory tract infection and pneumonia. Hypertrophic cardiomyopathy and signs of congestive heart failure were identified in the single FIV-positive cat from Group 1 that died.

3.4. Bodyweight

The bodyweight at the time of death for the single FIV-positive cat from Group 1 cohort that died was 15% heavier than the bodyweight recorded at enrolment (Fig. 2). The 17 FIV-positive cats from Group 2 that died lost a median of 51.3% of their bodyweight between enrolment (median 3.9 kg, range 2.27–5.54 kg) and death (median 1.9 kg, range 1.45–3.72 kg; $P < 0.0005$), over a median time span of 15 months (range 1.6–20 months). Those cats lost a median of 10.9% of their bodyweight between enrolment and 3 months before death (range −26.3% to +9.3%); median bodyweight loss in the last 3 months of life was 29.5% (range −55.4% to −8.9%).

Table 4. Organs with neoplastic lymphocyte infiltrates among FIV-positive cats diagnosed with lymphoma at post-mortem.

Cat #	Bone marrow	Lymph node	Spleen	Kidney	Liver	Jejunum	Tongue	Trachea	Heart	Eye
M3										+
M5	+	+	+							
M10	+	+								
M11	+	+	+	+	+	+	+	+	+	
M31	+									
M41	+	+	+							
M46	+									
M48	+	+								
M49	+									

None of the FIV-negative cats died during the study period.

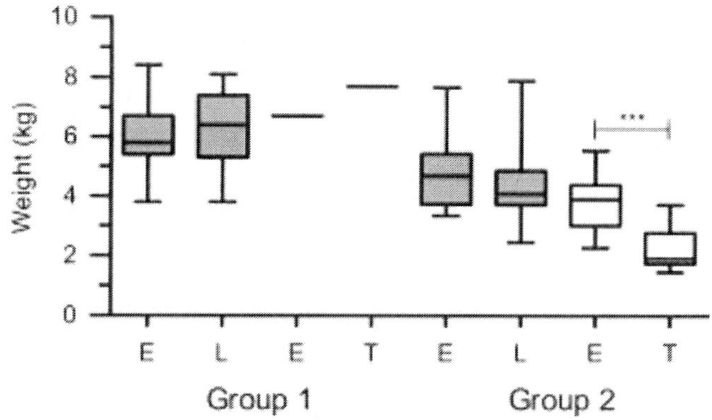

Figure 2. Bodyweights of FIV-positive cats from Group 1 (n = 17) and Group 2 (n = 27). E, enrolment weight; L, weight at the end of the study period; T, terminal weight. Grey boxes denote cats alive at the end of the 22-month study period; White boxes denote cats dead at the end of the 22-month study period. (The central bar in the box denotes the median, while the top and bottom of the box represent the 75th and 25th centiles, respectively. The upper and lower bars represent the 95th and 5th centiles, respectively. ***$P < 0.0005$.)

The 16 FIV-positive cats from Group 1 that survived for the entire study period gained a median of 7.6% bodyweight over the 12 months following enrolment (enrolment, median 5.9 kg, range 3.8–8.4 kg; 12 months later 6.35 kg, range 3.9–8.1 kg; $P = 0.2089$). In contrast, the 10 FIV positive cats from Group 2 that remained alive lost a median of 12.8% of their bodyweight over the 12 months following enrolment (enrolment, median 4.68 kg, range 3.34–7.67 kg; 12 months later, median 4.08 kg, range 2.45–7.89 kg; $P = 0.2324$).

FIV-negative cats from Group 1 for which bodyweight data were available (n = 16) gained a median 16.8% bodyweight between enrolment (median 5.35 kg, range 4–8.5 kg) and 12 months later (median 6.25 kg, range 4.1–8 kg; $P = 0.0145$). Similarly, FIV negative cats from Memphis for which bodyweight data were available (n = 24) gained a median of 24.5% bodyweight between enrolment (median, 4.9 kg, range 2.7–9 kg) and 12 months later (median 6.1 kg, range 2.6–9.9 kg; P = 0.0022).

Available bodyweight data from Group 2 demonstrated that enrolment weights for FIV-positive cats from Group 2 that died were significantly lower than those for matched FIV-negative cats (n = 17; FIV-positive, median 3.9 kg, range 2.27–5.54; FIV-negative, median 5.6 kg, range 2.7–6.8 kg; $P = 0.0046$). Bodyweights were not statistically different when FIV-positive survivors from Group 2 and matching FIV-negative cats were compared at enrolment ($n = 10$; FIV-positive, median 5.12 kg, range 3.34–7.68 kg; FIV-negative median 5.8 kg, range 4.6–9.0 kg; $P = 0.1641$) and 12 months later (FIV-positive, median 4.63 kg, range 3.63–8.27 kg; FIV-negative, median 6.2 kg, range 4.9–9.9 kg;$P = 0.0977$).

3.5. CD4+ T cell counts and CD4:CD8 ratio

Fig. 3 illustrates the CD4:CD8 ratios obtained over 12 months for the FIV-positive cats from Group 2 with reference values acquired from FIV-negative cats classified as healthy from the same geographic location. The difference between FIV-positive (median 0.89, range 0.19–2.24) and FIV-negative cats (median 1.71, range 1.03–2.62) was statistically significant at the time of enrolment ($P < 0.0001$) and 12 months later (medians 0.69, range 0.11–1.54 and 1.48, range 0.88–2.77 for FIV-positive and FIV-negative cats, respectively; $P < 0.0001$). The CD4:CD8 ratio in surviving FIV-positive cats was maintained over 18 months at a consistently low level (median values: 0.89, 0.73, 0.69, and 0.81 for each time point).

CD4:CD8 ratio at the time of enrolment ($P < 0.0005$) was lower in the FIV-positive cats from Group 1 (median 0.77, range 0.08–1.27) than in matched FIV-negative cats (median 2.18, range 1.23–6.42; Fig. 3). CD4:CD8 ratios in the FIV-positive cats from Group 1 were maintained at a consistently low level over the 12 months following enrolment (median 0.65, range 0.2–2.55) compared to age- and sex-matched FIV-negative cats (median 1.8, range 1.4–3.93; $P = 0.0054$). The CD4:CD8 ratios at the time of enrolment in the FIV-positive cats from Group 1 were lower (median 0.77, range 0.08–1.27) than those of the FIV-positive cats from Group 2 (median 0.89, range 0.19–2.23;Fig. 3).

There were no statistically significant differences when the CD4:CD8 ratio was compared between cats classified as healthy and not healthy at enrolment at either group, or for all FIV-positive cats as a single group (Group 1, $P = 0.97$; Group 2, $P = 0.84$; all FIV-positive cats, $P = 0.99$; Fig. 4).

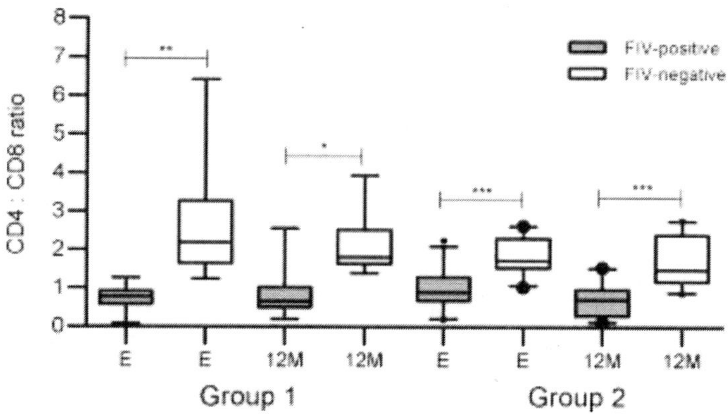

Figure 3. CD4:CD8 ratios for FIV-positive cats from Group 1 and Group 2 (in grey) and age- and sex-matched FIV-negative cats from the same geographic location (in white). For Group 1, data represent 17 FIV-positive cats and 17 age- and sex-matched FIV-negative cats at enrolment (E) and 16 surviving FIV-positive and 16 FIV-negative cats at 12 months after enrolment (12M). In Group 2 at enrolment, data were collected from 27 FIV-positive cats and 27 FIV-negative cats; 12 months after enrolment, data were collected from 21 surviving FIV-positive cats and 21 FIV-negative cats. (The central bar in the box denotes the median, while the top and bottom of the box represent the 75th and 25th centiles, respectively. The upper and lower bars represent the 95th and 5th centiles, respectively. $*P < 0.01$; $**P < 0.001$; $***P < 0.0001$.)

To determine the kinetics of the CD4:CD8 ratio inversion over the course of infection, we compared absolute CD4+ and CD8+ T lymphocyte numbers from 17 cats from Group 2 that died during the study (Fig. 5). The decrease in CD4+ T cell count was statistically significant when enrolment results (median 0.41 K/μL, range 0.13–1.38 K/μL) were compared with terminal results (median 0.18 K/μL, range 0.04–1.32 K/μL; $P = 0.0017$), but this was not the case for the decrease in CD8+ T cell count from enrolment (median 0.44 K/μL, range 0.18–1.49 K/μL) to terminal disease (median 0.38 K/μL, range 0.06–1.68 K/μL; $P = 0.35$). Therefore, the decrease in the CD4:CD8 ratio was attributed to a reduction in the CD4+ T cell count. Further analysis of total lymphocyte counts in deceased cats showed that the decrease in absolute lymphocyte numbers between enrolment (median 1.73 K/μL, range 1.15–4.03 K/μL) and terminal disease (median 1.25 K/μL, range 0.25–3.55 K/μL) reflected not only the loss of CD4+ T cells, but also depletion of CD21+ B cells (enrolment, median CD21+ B cell count 0.26

K/μL, range 0.12–0.89 K/μL; terminal specimens, median CD21+ B cell count 0.17 K/μL, range 0.04–0.37 K/μL; $P = 0.0061$).

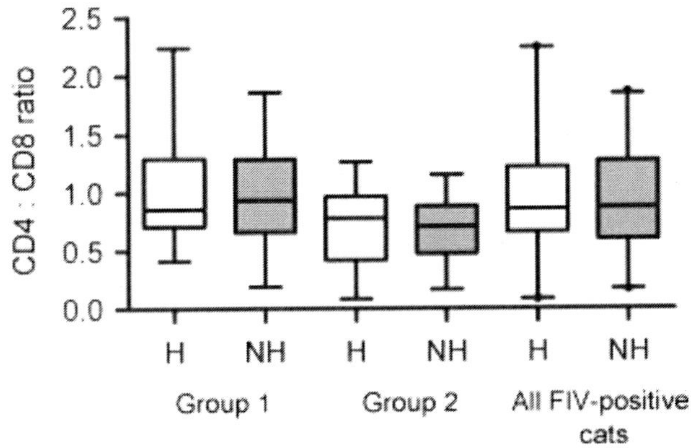

Figure 4. CD4:CD8 ratio at enrolment for FIV-positive cats from Group 1 ($n = 17$), Group 2 ($n = 27$) and all FIV-positive cats ($n = 44$). Cats were classified as healthy (H) or not healthy (NH) based on physical examination findings at the time of enrolment. (The central bar in the box denotes the median, while the top and bottom of the box represent the 75th and 25th centiles, respectively. The upper and lower bars represent the 95th and 5th centiles, respectively.)

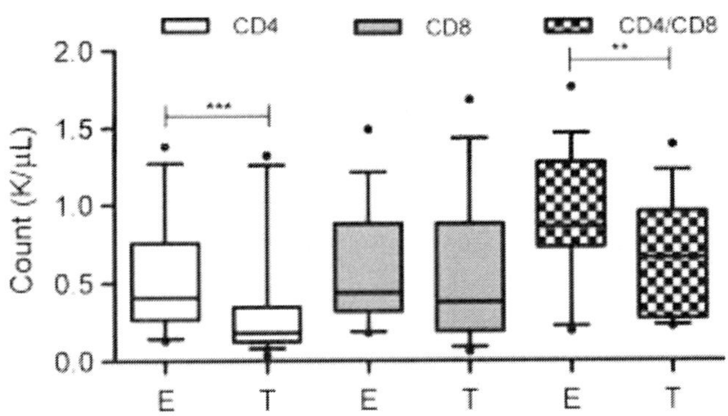

Figure 5. CD4+ and CD8+ T cell count and CD4:CD8 ratio at enrolment (E) and terminally (T) from the FIV-positive cats from Group 2 that died during the study ($n = 17$). (The central bar in the box denotes the median, while the top and bottom of the box represent the 75th and 25th centiles, respectively. The upper and lower bars represent the 95th and 5th centiles, respectively. $^{**}P = 0.0034$; $^{***}P = 0.0017$.)

3.6. FIV load (FIV genomes/mL blood)

The FIV load at enrolment was significantly lower in Group 1 (median, 741 genomes/mL blood; range 43–18,796 genomes/mL blood) than Group 2 (median, 50,003 genomes/mL blood; range, 2540–3,792,000 genomes/mL blood; $P < 0.0001$; Fig. 6). There were no statistically significant differences in FIV load at enrolment when healthy and not healthy cats were compared in Group 1 (healthy, $n = 8$, median 717 genomes/mL blood; range, 43–18,796 genomes/mL blood; not healthy, $n = 7$, median 767 genomes/mL blood; range, 395–18,796 genomes/mL blood; $P > 0.05$) or Group 2 (healthy, $n = 11$, median 64,843 genomes/mL blood; range, 2540–1,667,000 genomes/mL blood; not healthy, $n = 15$, median 41,387 genomes/mL blood; range, 13,423–3,792,000 genomes/mL blood; $P > 0.05$). In Group 2, there were no significant differences in FIV load at enrolment when the following comparisons were made: cats still alive vs. cats that had died by the end of the study period (alive, $n = 9$, median 31,395 genomes/mL blood; range, 2540–41,8671 genomes/mL blood; dead, $n = 17$, median 176,551 genomes/mL blood; range, 7724–3,792,000 genomes/mL blood; $P > 0.05$); cats that had died by the end of the study period and were classified at enrolment as either healthy or not healthy (healthy, $n = 6$, median 241,757 genomes/mL blood, range, 7724–1,667,000 genomes/mL blood; not healthy, $n = 11$, median 176,551 genomes/mL blood; range, 13,423–3,792,000 genomes/mL blood; $P > 0.05$); and cats that were still alive by the end of the study period and were classified at enrolment as either healthy or not healthy (healthy, $n = 5$, median 56,867 genomes/mL blood, range, 2540–418,671 genomes/mL blood; not healthy, $n = 4$, median 27,771 genomes/mL blood, range 16,176–43,138 genomes/mL blood; $P > 0.05$).

Statistical correlations between FIV load and CD4+ T cells, CD8+ T cells and the CD4:CD8 ratio were investigated for each group. There were statistically significant, or close to significant, negative correlations between the CD4:CD8 ratio and FIV load in Group 1 ($r = -0.5022$; $P = 0.0564$) and in Group 2 ($r = -0.5479$; $P = 0.0038$; Fig. 7).

3.7. Phylogenetic classification of FIV envelope genes

Full-length viral envelope (*env*) gene sequences from FIV-positive cats were examined ($n = 355$). In Group 1, clade B *env* viruses and clade A/B recombinant *env* viruses each accounted for 50% of infections. In Group

2, 69% of cats were infected with viruses bearing clade B *env*, while 31% of cats were infected with recombinant *env* viruses (Supplementary Table 3).

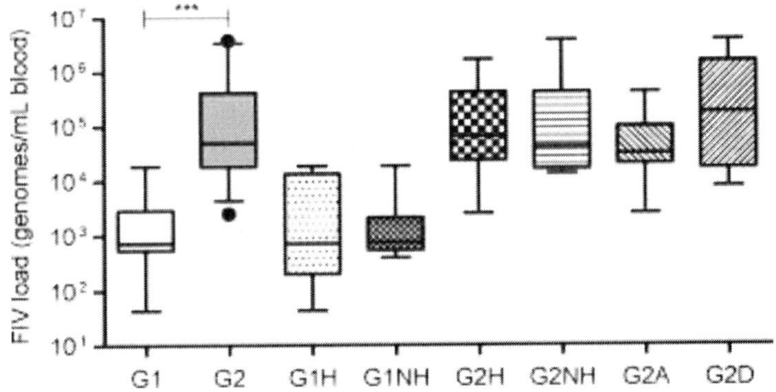

Figure 6. FIV load at the time of enrolment (genomes/mL blood). Statistical comparisons were made between pairs of classifications across the graph. G1, Group 1 ($n = 15$); G2, Group 2 ($n = 26$); H, healthy on the day of enrolment (no abnormalities detected on a physical examination; Group 1, $n = 8$, Group 2,$n = 11$); NH, not healthy on the day of enrolment (at least one abnormality detected on a physical examination; Group 1, $n = 7$, Group 2, $n = 15$); A, alive at the end of the 22-month study period (Group 2,$n = 9$); D, cats dead at the end of the 22-month study period (Group 2, $n = 17$). ***$P < 0.0001$.

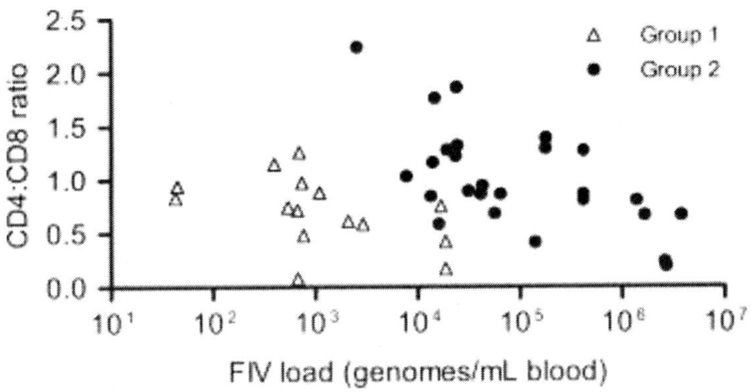

Figure 7. Correlation between FIV load (FIV genomes/mL blood) and CD4:CD8 ratio for FIV-positive cats from Group 1 ($n = 15$; $r = -0.5022$, $P = 0.0564$) and Group 2 ($n = 26$; $r = -0.5479$, $P = 0.0038$).

4. DISCUSSION

Here we describe contrasting clinical outcomes of naturally acquired FIV infection, demonstrating significant differences between cats enrolled from two separate cohorts. The relationship between FIV infection and clinical disease is unclear, with some studies reporting that infected cats are at increased risk of morbidity and mortality (Muirden, 2002), while others report no such association (Akhtardanesh et al., 2010 and Liem et al., 2013). Considering all 44 FIV-positive cats, there was no relationship between mortality and morbidity and FIV infection; however, this was not the case when the two cohorts were examined separately. Our observations in Group 2 indicated that living conditions might have played a role in the more rapid disease onset. In contrast, cats from Group 1 were generally housed in single cat households and remained alive and in relatively good health over the study period. Therefore, it appears that FIV infection is more likely to progress in cats kept in crowded shelter conditions compared to those living in spacious environments. As a corollary, our results suggest that the FIV-positive cats in our study that lived in single or dual cat households remained free of clinical signs of illness over the study period; this can most likely be attributed to lower risks of exposure to opportunistic pathogens and reduced levels of environmental stress.

FIV infected cats are at greater risk of opportunistic infections and neoplasia (Hartmann, 2012). Regular assessment of their health status is essential and, according to European Advisory Board on Cat Diseases guidelines (Hosie et al., 2009), health checks are recommended every 6 months. The short time to disease progression leading to the deaths of 63% of cats in Group 2 within the 22-month follow-up period was unexpectedly high, especially in light of previously published studies that were unable to demonstrate a difference in lifespan between FIV-positive and FIV-negative cats (Addie et al., 2000 and Ravi et al., 2010). The health status of cats from Group 1 was similar to that of cats in a household in which FIV infection did not reduce life expectancy over a period of 2 years (Addie et al., 2000). There are several clinical abnormalities associated with FIV infection, with gingivostomatitis being the most common. Although one study (Quimby et al., 2008) did not observe an association between FIV infection and gingivostomatitis, the high prevalence observed here was in agreement with previous reports (Bandecchi et al., 1992, Hofmann-Lehmann et al., 1995, Hosie et al., 2009 and Knotek et al., 1999). As gingivostomatitis is rare in SPF cats experimentally infected with FIV, co-infections with other infectious agents, such as feline calicivirus, might have contributed to this syndrome in naturally infected cats (Reubel et al., 1994 and Tenorio et al., 1991). It is perhaps relevant that severe acute on chronic upper respiratory tract disease was diagnosed in four cats from Group 2 at post-mortem. Other abnormalities previously attributed to FIV-induced dysregulation and impairment of immune surveillance include dermatitis, ocular disease, renal insufficiency, lower urinary tract infections and other opportunistic infections (Barlough et al., 1991 and Reche et al., 2010). Although neurological abnormalities have been reported in natural and experimental, acute and chronic FIV infections (Abramo et al., 1995, Dow et al., 1990, Podell et al., 1997 and Podell et al., 1999), neither behavioural nor motor abnormalities were observed in our study population.

The mortality rate of 63% in the FIV-positive cats from Group 2 was markedly higher than the FIV-positive cats from Group 1, in which only one cat died during the study period (6%). Post-mortem examinations identified several pathological changes in the FIV-infected cats. Although FIV infection is generally associated with immunosuppression, immune hyperstimulation is evident during the early stages of infection, including B cell activation, manifested as polyclonal hypergammaglobulinemia (Hopper et al., 1989 and Takano et al., 2012), expansion of CD8+ T cells with an activated phenotype and subsequent CD8+ T lymphocytosis (Beatty et al., 1996 and Willett et al., 1993). Lymph node follicular hyperplasia and dysplasia was documented post-mortem in this study, confirming the findings of other studies of acute and terminal stage disease (Matsumura et al., 1993, Shelton et al., 1989 and Yamamoto et al., 1988). In addition, reactive changes in the bone marrow, observed in 56% of cats in this study, have been documented previously in HIV and FIV infections (Bain et al., 2009, Breuer et al., 1998 and Geller et al., 1985).

Lymphoma was the most common condition diagnosed at post-mortem in Group 2 (9/16 cats; 56%), which raises questions regarding potential interactions between virus, host, environment, and other factors that could have resulted in a predisposition to neoplasia. FIV strain variability was examined, since the Memphis FIV strains (Group 2) might have had greater oncogenic potential than the strains infecting the Chicago cohort (Group 1). However, a comparison of full-length *env* gene and long terminal repeat sequences revealed no significant differences (unpublished observation). It is possible that the local cluster of lymphoma in Group 2 was associated with another infectious agent, such as FeLV, but this seemed unlikely since FIV/FeLV Snap test (IDEXX) results were negative for FeLV antigen. However, we cannot exclude the possibility of regressive FeLV infection (Hofmann-Lehmann et al., 2001), since PCR testing to detect integrated proviral DNA in bone marrow was not performed. Breed was unlikely to play a role, as the cats from Group 2 diagnosed with lymphoma and those from Group 1 were all classified as domestic shorthairs and came from a variety of sources within their geographic area. The shelter environment in which the cats from Group 2 lived was likely to play a role in the development of the various clinical manifestations of FIV, although exposure to potential carcinogens cannot be ruled out.

Marked weight loss between enrolment and terminal measurement was evident among the cats that died over the study period, whereas healthy FIV-infected cats maintained relatively stable bodyweights (Fig. 2). Although monitoring bodyweight regularly can be a valuable tool in the care of FIV-positive cats and appeared to have some prognostic value in our study of FIV-positive cats, declining bodyweight is a relatively non-specific finding, as it is not unusual to observe weight loss during the progression of various other diseases.

The hallmark of FIV infection is an inverted CD4:CD8 ratio, which, as demonstrated here and elsewhere (Hoffmann-Fezer et al., 1992, Novotney et al., 1990 and Tompkins et al., 1991), is largely the consequence of a loss of CD4+ T cells. Here we provide further evidence that FIV-positive cats had significantly lower CD4:CD8 ratios than those for FIV-negative cats over the 22-month study period in each of the cohorts (Fig. 3). We also observed a statistically significant decline in the CD4:CD8 ratio between the time of enrolment and the terminal sampling in those cats that died over the study period (Fig. 5). However, there was not a significant difference in median CD4:CD8 ratio at enrolment when cats classified as either healthy or not healthy were compared (Fig. 4). Both groups of cats had relatively low CD4:CD8 ratios with median values less than 1, regardless of their health status. Indeed, the cat that had the lowest CD4+ T lymphocyte count (0.09 K/μL) and the lowest CD4:CD8 ratio (0.08) at enrolment was classified as healthy and is still alive. Similar observations have been reported previously in FIV (Novotney et al., 1990) and HIV infections (Levy, 1988), where seropositive individuals remained asymptomatic despite significant loss of CD4+ T lymphocytes. The variability in CD4:CD8 ratios among healthy and not healthy cats in this study population also limited their utility in the prediction of disease progression.

Extrapolating from HIV infection, the plasma viral load has been postulated as a potential prognostic indicator for FIV infection. A study focusing on disease progression in SPF cats experimentally infected with two strains of FIV of different virulence reported a correlation between disease progression and increased plasma viral load (Diehl and Hoover,

1995). Similarly, the relationship between plasma viral load, disease progression and survival time was found in a study of naturally infected cats in Japan (Goto et al., 2002). In the present study, the median FIV load at the time of enrolment from Group 2 was 67-fold higher than that from Group 1 ($P < 0.0001$) and there was a marked difference in health status and mortality rate between groups over the following 22 months. It is not possible to discern whether an increased viral load was a cause or a consequence of poor health status, since that would have required knowledge of viral load kinetics and changes in health status from the time of initial infection in the FIV-infected 'healthy' and 'not healthy' cats; in this prospective study the data were collected from the time of enrolment. However, statistically significant differences in FIV load between cats from Group 2 classified as 'healthy' and 'not healthy' and cats that remained alive or died during the study were not apparent. Observation of these parameters over longer periods of time, preferably with larger group sizes, would be required to fully validate their utility in predicting disease progression.

It has been hypothesized that FIV clade B viruses are evolutionarily older and more host-adapted, hence less pathogenic, than those of clade A (Sodora et al., 1994). One possible explanation for the high morbidity and mortality in Group 2 is that those cats were infected with highly virulent strains of FIV. However, this hypothesis was not supported by the phylogenetic analyses performed, as cats from both groups were infected with comparable numbers of clade B and recombinant A/B viruses. Furthermore, sequences isolated from serial 6-monthly blood collections demonstrated that highly monophyletic groups were formed in 96% of the cats from Group 2, suggesting that viral isolates were not transmitted between cats despite unrestricted cohabitation (Bęczkowski et al., 2014). Although we cannot exclude the possibility that cats from Memphis were infected with more virulent FIV isolates, based on thorough phylogenetic analyses of full-length *env*sequences, this seems unlikely.

A limitation of this study is the restriction of enrolments to two distinct geographic areas; the inclusion of cats from a wider range of locations could have increased the variability in virulence of the wild type FIVs examined and their possible contribution to the onset of disease. Additionally, for the purposes of this study of naturally infected cats either adopted from shelters or rescued, it was necessary to estimate the ages of the cats and to use the date that the cat first tested FIV-positive since the precise dates of infection were not known. It has been reported previously that age at time of first exposure to FIV can influence the course of infection (George et al., 1993) and can determine the outcome of disease (Akhtardanesh et al., 2010). From a biological and behavioural perspective, the age of approximately 1–2 years old at the time of first diagnosis reported in this study is consistent with young cats engaging in territorial fights, especially if they are entire. The cats in our study were originally adopted from shelters and many were entire and had outdoor access at the time of shelter intake, both known lifestyle risk factors for FIV (Levy et al., 2006). In our study population, infection was more common in male cats, confirming the results of previous studies (Grindem et al., 1989, Hosie et al., 1989, Ishida et al., 1989,Levy et al., 2006 and Yamamoto et al., 1989). Additionally, FIV status and multi-cat housing status could have affected the health of the cats in our study; it was not possible to discern whether they acted independently or as cofactors. FIV-negative cats matched to enrolled FIV-positive cats on housing status, as well as age and sex, were not available during the study enrolment period. While an extra layer of matching based on housing status would have been ideal, it was not achievable in this real-world study.

Based on the relatively large number of cats enrolled and the striking clinical differences observed between the two cohorts, we conclude that keeping FIV-infected cats in overcrowded conditions can have a significant impact on the risk of disease progression, particularly in cats which already have their immune systems compromised by FIV infection. In contrast, FIV-positive cats remained in relatively good health when living in stable, single cat households. Clinical signs of disease were not observed at 6-monthly physical examinations over a 24-month period of the cats from Group 1 that were housed in single cat households. The average duration of the 'asymptomatic' phase of infection remains unknown and could only be estimated following observations over a longer period. In this study, the selection of study participants from different locations and housing conditions minimized potential selection bias. It is apparent that the conclusions of this study would have been very different indeed had efforts been focused on only one of the cohorts presented herein.

CONFLICT OF INTEREST

The authors declare that they have no competing interests.

ACKNOWLEDGEMENTS

This study was supported by The Wellcome Trust and the Maddie's Fund. The Purdue Maddie's Shelter Medicine Programme is underwritten by a grant from Maddie's Fund, The Pet Rescue Foundation (www.maddiesfund.org), helping to fund the creation of a no-kill nation. We thank Kristen Hall CVT, Dr. Jui Ming Lin, Dr. Christian Leutenegger, PAWS Chicago, Drennan Animal Hospital, the Fitzhugh B. Crews FIV Cat Sanctuary and participating cat owners for their assistance with the study. We thank the IDEXX Corporation (Westbrook, MN and West Sacramento, CA) for providing the FIV load and flow cytometry results.

REFERENCES

1. Abramo, F., Bo, S., Canese, M.G., Poli, A., 1995. Regional distribution of lesions in the central nervous system of cats infected with feline immunodeficiency virus. AIDS Res. Hum. Retrovir. 11, 1247–1253.

2. Addie, D.D., Dennis, J.M., Toth, S., Callanan, J.J., Reid, S., Jarrett, O., 2000. Long-term impact on a closed household of pet cats of natural infection with feline coronavirus, feline leukaemia virus and feline immunodeficiency virus. Vet. Rec. 146, 419–424.

3. Akhtardanesh, B., Ziaali, N., Sharifi, H., Rezaei, S., 2010. Feline immunodeficiency virus, feline leukemia virus and Toxoplasma gondii in stray and household cats in Kerman – Iran: seroprevalence and correlation with clinical and laboratory findings. Res. Vet. Sci. 89, 306–310.

4. Bain, B.J., Clark, D.M., Wilkins, B.S., 2009. Infection and reactive changes. In: Bone Marrow Pathology. Wiley-Blackwell, , pp. 100–165.

5. Bandecchi, P., Matteucci, D., Baldinotti, F., Guidi, G., Abramo, F., Tozzini, F., Bendinelli, M., 1992. Prevalence of feline immunodeficiency virus and other retroviral infections in sick cats in Italy. Vet. Immunol. Immunopathol. 31, 337–345

6. Barlough, J.E., Ackley, C.D., George, J.W., Levy, N., Acevedo, R., Moore, P.F., Rideout, B.A., Cooper, M.D., Pedersen, N.C., 1991. Acquired immune dysfunction in cats with experimentally induced feline immunodefi- ciency virus infection: comparison of short-term and long-term infections. J. Acquir. Immune Defic. Syndr. 4, 219–227.

7. Beatty, J.A., Willett, B.J., Gault, E.A., Jarrett, O., 1996. A longitudinal study of feline immunodeficiency virus-specific cytotoxic T lymphocytes in experimentally infected cats, using antigen-specific induction. J. Virol. 70, 6199–6206.

8. Bęczkowski, P.M., Hughes, J., Biek, R., Litster, A., Willett, B.J., Hosie, M.J., 2014. Feline immunodeficiency virus (FIV) env recombinants are common in natural infections. Retrovirology 11, 80.

9. Breuer, W., Stahr, K., Majzoub, M., Hermanns, W., 1998. Bone-marrow changes in infectious diseases and lymphohaemopoietic neoplasias in dogs and cats—a retrospective study. J. Comp. Pathol. 119, 57–66.

10. Churchill, D.R., 1997. Prognostic markers and surrogate markers of clinical progression in HIV infection. Int. J. STD AIDS 8, 552–556, quiz 557.

11. Diehl, L.J., Hoover, E.A., 1995. Disease progression correlates with plasma viral-RNA load in an accelerated FIV model. AIDS Res. Hum. Retrovir. 11, S93.

12. Dow, S.W., Poss, M.L., Hoover, E.A., 1990. Feline immunodeficiency virus: a neurotropic lentivirus. J. Acquir. Immune Defic. Syndr. 3, 658–668.

13. Efron, B., Halloran, E., Holmes, S., 1996. Bootstrap confidence levels for phylogenetic trees. Proc. Natl. Acad. Sci. U. S. A. 93, 13429–13434.

14. English, R.V., Nelson, P., Johnson, C.M., Nasisse, M., Tompkins, W.A., Tompkins, M.B., 1994. Development of clinical disease in cats experimentally infected with feline immunodeficiency virus. J. Infect. Dis. 170, 543–552.

15. Geller, S.A., Muller, R., Greenberg, M.L., Siegal, F.P., 1985. Acquired immunodeficiency syndrome distinctive features of bone marrow biopsies. Arch. Pathol. Lab. Med. 109, 138–141.

16. George, J.W., Pedersen, N.C., Higgins, J., 1993. The effect of age on the course of experimental feline immunodeficiency virus infection in cats. AIDS Res. Hum. Retrovir. 9, 897–905.

17. Goto, Y., Nishimura, Y., Baba, K., Mizuno, T., Endo, Y., Masuda, K., Ohno, K., Tsujimoto, H., 2002. Association of plasma viral RNA load with prognosis in cats naturally infected with feline immunodeficiency virus. J. Virol. 76, 10079–10083.

18. Grindem, C.B., Corbett, W.T., Ammerman, B.E., Tomkins, M.T., 1989. Seroepidemiologic survey of feline immunodeficiency virus infection in cats of Wake County North Carolina. J. Am. Vet. Med. Assoc. 194, 226–228.

19. Hartmann, K., 1998. Feline immunodeficiency virus infection: an overview. Vet. J. 155, 123–137.

20. Hartmann, K., 2011. Clinical aspects of feline immunodeficiency and feline leukemia virus infection. Vet. Immunol. Immunopathol. 143, 190–201.

21. Hartmann, K., 2012. Clinical aspects of feline retroviruses: a review. Viruses 4, 2684–2710.

22. Hoffmann-Fezer, G., Thum, J., Ackley, C., Herbold, M., Mysliwietz, J., Thefeld, S., Hartmann, K., Kraft, W., 1992. Decline in CD4+ cell numbers in cats with naturally acquired feline immunodeficiency virus infection. J. Virol. 66, 1484–1488.

23. Hofmann-Lehmann, R., Holznagel, E., Aubert, A., Bauer-Pham, K., Lutz, H., 1995. FIV vaccine studies II. Clinical findings, hematological changes and kinetics of blood lymphocyte subsets. Vet. Immunol. Immunopathol. 46, 115–125.

24. Hofmann-Lehmann, R., Holznagel, E., Ossent, P., Lutz, H., 1997. Parameters of disease progression in long-term experimental feline retrovirus (feline immunodeficiency virus and feline leukemia virus) infections: hematology, clinical chemistry, and lymphocyte subsets. Clin. Diagn. Lab. Immunol. 4, 33–42.

25. Hofmann-Lehmann, R., Huder, J.B., Gruber, S., Boretti, F., Sigrist, B., Lutz, H., 2001. Feline leukemia provirus load during the course of experimental infection and in naturally infected cats. J. Gen. Virol. 82, 1589–1596.

26. Hopper, C.D., Sparkes, A.H., Gruffydd-Jones, T.J., Crispin, S.M., Muir, P., Harbour, D.A., Stokes, C.R., 1989. Clinical and laboratory findings in cats infected with feline immunodeficiency virus. Vet. Rec. 125, 341–346.

27. Hosie, M.J., Robertson, C., Jarrett, O., 1989. Prevalence of feline leukaemia virus and antibodies to feline immunodeficiency virus in cats in the United Kingdom. Vet. Rec. 125, 293–297.

28. Hosie, M.J., Addie, D., Bela´k, S., Boucraut-Baralon, C., Egberink, H., Frymus, T., Gruffydd-Jones, T., Hartmann, K., Lloret, A., Lutz, H., Marsilio, F., Pennisi, M.G., Radford, A.D., Thiry, E., Truyen, U., Horzinek, M.C., 2009. Feline immunodeficiency ABCD guidelines on prevention and management. J. Feline Med. Surg. 11, 575–584.

29. Ishida, T., Washizu, T., Toriyabe, K., Motoyoshi, S., Tomoda, I., Pedersen, N.C., 1989. Feline immunodeficiency virus infection in cats of Japan. J. Am. Vet. Med. Assoc. 194, 221–225.

30. Katzenstein, T.L., 2003. Molecular biological assessment methods and understanding the course of the HIV infection. APMIS Suppl. 1–37.

31. Knotek, Z., Hajkova, P., Svoboda, M., Toman, M., Raska, V., 1999. Epidemiology of feline leukaemia and feline immunodeficiency virus infections in the Czech Republic. Zent. Veterinarmed. B 46, 665–671.

32. Kohmoto, M., Uetsuka, K., Ikeda, Y., Inoshima, Y., Shimojima, M., Sato, E., Inada, G., Toyosaki, T., Miyazawa, T., Doi, K., Mikami, T., 1998. Eightyear observation and comparative study of specific pathogen-free cats experimentally infected with feline immunodeficiency virus (FIV) subtypes A and B: terminal acquired immunodeficiency syndrome in a cat infected with FIV petaluma strain. J. Vet. Med. Sci. 60, 315–321.

33. Levy, J.A., 1988. Mysteries of HIV: challenges for therapy and prevention. Nature 333, 519–522.

34. Levy, J.K., Scott, H.M., Lachtara, J.L., Crawford, P.C., 2006. Seroprevalence of feline leukemia virus and feline immunodeficiency virus infection among cats in North America and risk factors for seropositivity. J. Am. Vet. Med. Assoc. 228, 371–376.

35. Liem, B.P., Dhand, N.K., Pepper, A.E., Barrs, V.R., Beatty, J.A., 2013. Clinical findings and survival in cats naturally infected with feline immunodeficiency virus. J. Vet. Intern. Med. 27, 798–805.

36. Lin, J., Litster, A., 2013. Fluorescence flow cytometry methodology to exclude platelet aggregate interference when measuring feline CD4 and CD8 lymphocyte counts. Vet. J. 198, 275–278.

37. Lobprise, H.B., 2007. Oral exam and charting. In: Blackwell's Five Minute Veterinary Consult Clinical Companion: Small Animal Dentistry. first ed. Blackwell Publishing Ltd., Ames, IA, USA. Matsumura, S., Ishida, T., Washizu, T., Tomoda, I., Nagata, S., Chiba, J., Kurata, T., 1993. Pathologic features of acquired immunodeficiencylike syndrome in cats experimentally infected with feline immunodeficiency virus. J. Vet. Med. Sci. 55, 387–394.

38. Muirden, A., 2002. Prevalence of feline leukaemia virus and antibodies to feline immunodeficiency virus and feline coronavirus in stray cats sent to an RSPCA hospital. Vet. Rec. 150, 621–625.

39. Novotney, C., English, R.V., Housman, J., Davidson, M.G., Nasisse, M.P., Jeng, C.R., Davis, W.C., Tompkins, M.B., 1990. Lymphocyte population changes in cats naturally infected with feline immunodeficiency virus. AIDS 4, 1213–1218.

40. Okonji, J.A., Zeh, C., Weidle, P.J., Williamson, J., Akoth, B., Masaba, R., Fowler, M.G., Thomas, T.K., 2012. CD4, viral load response and adherence among antiretroviral-naive breastfeeding women receiving triple antiretroviral prophylaxis for prevention of mother-to-child transmission of HIV in Kisumu, Kenya. J. Acquir. Immune Defic. Syndr. 61 (2) 249–257.

41. Pedersen, N.C., Leutenegger, C.M., Woo, J., Higgins, J., 2001. Virulence differences between two field isolates of feline immunodeficiency virus (FIV-apetaluma and FIV-cpgammar) in young adult specific pathogen free cats. Vet. Immunol. Immunopathol. 79, 53–67.

42. Podell, M., Hayes, K., Oglesbee, M., Mathes, L., 1997. Progressive encephalopathy associated with CD4/CD8 inversion in adult FIVinfected cats. J. Acquir. Immune Defic. Syndr. Hum. Retrovirol 15, 332–340.

43. Podell, M., Maruyama, K., Smith, M., Hayes, K.A., Buck, W.R., Ruehlmann, D.S., Mathes, L.E., 1999. Frontal lobe neuronal injury correlates to altered function in FIV-infected cats. J. Acquir. Immune Defic. Syndr. Hum. Retrovirol 22, 10–18.

44. Posada, D., 2008. jMODELTEST: phylogenetic model averaging. Mol. Biol. Evol. 25, 1253–1256. Quimby, J.M., Elston, T., Hawley, J., Brewer, M., Miller, A., Lappin, M.R., 2008. Evaluation of the association of Bartonella species, feline herpesvirus 1, feline calicivirus, feline leukemia virus and feline immunodeficiency virus with chronic feline gingivostomatitis. J. Feline Med. Surg. 10, 66–72.

45. Ravi, M., Wobeser, G.A., Taylor, S.M., Jackson, M.L., 2010. Naturally acquired feline immunodeficiency virus (FIV) infection in cats from Western Canada: prevalence, disease associations, and survival analysis. Can. Vet. J. 51, 271–276.

46. Reche Jr., A., Daniel, A.G.T., Lazaro Strauss, T.C.P., Taborda, C.P., Vieira Marques, S.A., Haipek, K., Oliveira, L.J., Monteiro, J.M., Kfoury Jr., J.R., 2010. Cutaneous mycoflora and CD4:CD8 ratio of cats infected with feline immunodeficiency virus. J. Feline Med. Surg. 12, 355–358.

47. Reubel, G.H., George, J.W., Higgins, J., Pedersen, N.C., 1994. Effect of chronic feline immunodeficiency virus infection on experimental feline calicivirus-induced disease. Vet. Microbiol. 39, 335–351.

48. Shaunak, S., Teo, I., 2003. Monitoring HIV disease with new and clinically useful surrogate markers. Curr. Opin. Infect. Dis. 16, 581–586.

49. Shelton, G.H., Abkowitz, J.L., Linenberger, M.L., Russell, R.G., Grant, C.K., 1989. Chronic leukopenia associated with feline immunodeficiency virus infection in a cat. J. Am. Vet. Med. Assoc. 194, 253–255.

50. Smith, C.L., Stein, G.E., 2002. Viral load as a surrogate end point in HIV disease. Ann. Pharmacother. 36, 280–287.

51. Sodora, D.L., Shpaer, E.G., Kitchell, B.E., Dow, S.W., Hoover, E.A., Mullins, J.I., 1994. Identification of three feline immunodeficiency virus (FIV) env gene subtypes and comparison of the FIV and human immunodeficiency virus type 1 evolutionary patterns. J. Virol. 68, 2230–2238.

52. Takano, T., Hosoya, S., Shibao, A., Nagasaki, B., Yoshioka, H., Satoh, R., Hohdatsu, T., 2012. Comparative study of the plasma globulin level CD21() B-cell counts and FOXP3 mRNA expression level in CD4(+) Tcells for different clinical stages of feline immunodeficiency virus infected cats. Res. Vet. Sci. 92, 157–161.

53. Tamura, K., Peterson, D., Peterson, N., Stecher, G., Nei, M., Kumar, S., 2011. Mega5: molecular evolutionary genetics analysis using maximum likelihood, evolutionary distance, and maximum parsimony methods. Mol. Biol. Evol. 28, 2731–2739.

54. Tenorio, A.P., Franti, C.E., Madewell, B.R., Pedersen, N.C., 1991. Chronic oral infections of cats and their relationship to persistent oral carriage of feline calici-, immunodeficiency, or leukemia viruses. Vet. Immunol. Immunopathol. 29, 1–14.

55. Tompkins, M.B., Nelson, P.D., English, R.V., Novotney, C., 1991. Early events in the immunopathogenesis of feline retrovirus infections. J. Am. Vet. Med. Assoc. 199, 1311–1315.

56. Willett, B.J., Hosie, M.J., Callanan, J.J., Neil, J.C., Jarrett, O., 1993. Infection with feline immunodeficiency virus is followed by the rapid expansion of a CD8+ lymphocyte subset. Immunology 78, 1–6.

57. Yamamoto, J.K., Sparger, E., Ho, E.W., Andersen, P.R., O'Connor, T.P., Mandell, C.P., Lowenstine, L., Munn, R., Pedersen, N.C., 1988. Pathogenesis of experimentally induced feline immunodeficiency virus infection in cats. Am. J. Vet. Res. 49, 1246–1258.

58. Yamamoto, J.K., Hansen, H., Ho, E.W., Morishita, T.Y., Okuda, T., Sawa, T.R., Nakamura, R.M., Pedersen, N.C., 1989. Epidemiologic and clinical aspects of feline immunodeficiency virus infection in cats from the continental United States and Canada and possible mode of transmission. J. Am. Vet. Med. Assoc. 194, 213–220.

Index

N

P

R

S

T

V

Z